D0148394

ACS SYMPOSIUM SERIES **381**

The Effects of Radiation on High-Technology Polymers

Elsa Reichmanis, EDITOR
AT&T Bell Laboratories

James H. O'Donnell, EDITOR
University of Queensland

Developed from a workshop sponsored
by the Division of Polymer Chemistry, Inc.,
of the American Chemical Society
and the Polymer Division
of the Royal Australian Chemical Institute,
Queensland, Australia
August 16–19, 1987

Bowling Green State University
Science Library

JAN 27 1989

SERIALS

American Chemical Society, Washington, DC 1989

Library of Congress Cataloging-in-Publication Data

The effects of radiation on high-technology polymers
   Elsa Reichmanis, editor, James H. O'Donnell, editor.

   p.   cm.—(ACS Symposium Series, ISSN 0097–6156; 381)

   Developed from a workshop sponsored by the Division
of Polymer Chemistry, Inc., of the American Chemical
Society and the Polymer Division of the Royal Australian
Chemical Institute, Queensland, Australia, August 16–19,
1987.

   Bibliography: p.

   Includes index.

   ISBN 0–8412–1558–8

   1. Polymers and polymerization—Effect of radiation
on—Congresses.

   I. Reichmanis, Elsa, 1953–   . II. O'Donnell, James H.
III. American Chemical Society. Division of Polymer
Chemistry. IV. Royal Australian Chemical Institute.
Polymer Division. V. Series.

QD381.9.R3E42 1989
620.1'9204228—dc19   88–39298
                                                 CIP

Copyright ᶜ 1989

American Chemical Society

All Rights Reserved. The appearance of the code at the bottom of the first page of each
chapter in this volume indicates the copyright owner's consent that reprographic copies
of the chapter may be made for personal or internal use or for the personal or internal
use of specific clients. This consent is given on the condition, however, that the copier
pay the stated per-copy fee through the Copyright Clearance Center, Inc., 27 Congress
Street, Salem, MA 01970, for copying beyond that permitted by Sections 107 or 108 of the
U.S. Copyright Law. This consent does not extend to copying or transmission by any
means—graphic or electronic—for any other purpose, such as for general distribution, for
advertising or promotional purposes, for creating a new collective work, for resale, or for
information storage and retrieval systems. The copying fee for each chapter is indicated in
the code at the bottom of the first page of the chapter.

The citation of trade names and/or names of manufacturers in this publication is not to
be construed as an endorsement or as approval by ACS of the commercial products or
services referenced herein; nor should the mere reference herein to any drawing,
specification, chemical process, or other data be regarded as a license or as a conveyance
of any right or permission to the holder, reader, or any other person or corporation, to
manufacture, reproduce, use, or sell any patented invention or copyrighted work that may
in any way be related thereto. Registered names, trademarks, etc., used in this publication,
even without specific indication thereof, are not to be considered unprotected by law.

PRINTED IN THE UNITED STATES OF AMERICA

ACS Symposium Series

M. Joan Comstock, *Series Editor*

1988 ACS Books Advisory Board

Paul S. Anderson
Merck Sharp & Dohme Research
  Laboratories

Harvey W. Blanch
University of California—Berkeley

Malcolm H. Chisholm
Indiana University

Alan Elzerman
Clemson University

John W. Finley
Nabisco Brands, Inc.

Natalie Foster
Lehigh University

Marye Anne Fox
The University of Texas—Austin

Roland F. Hirsch
U.S. Department of Energy

G. Wayne Ivie
USDA, Agricultural Research Service

Michael R. Ladisch
Purdue University

Vincent D. McGinniss
Battelle Columbus Laboratories

Daniel M. Quinn
University of Iowa

James C. Randall
Exxon Chemical Company

E. Reichmanis
AT&T Bell Laboratories

C. M. Roland
U.S. Naval Research Laboratory

W. D. Shults
Oak Ridge National Laboratory

Geoffrey K. Smith
Rohm & Haas Co.

Douglas B. Walters
National Institute of
  Environmental Health

Wendy A. Warr
Imperial Chemical Industries

Foreword

The ACS SYMPOSIUM SERIES was founded in 1974 to provide a medium for publishing symposia quickly in book form. The format of the Series parallels that of the continuing ADVANCES IN CHEMISTRY SERIES except that, in order to save time, the papers are not typeset but are reproduced as they are submitted by the authors in camera-ready form. Papers are reviewed under the supervision of the Editors with the assistance of the Series Advisory Board and are selected to maintain the integrity of the symposia; however, verbatim reproductions of previously published papers are not accepted. Both reviews and reports of research are acceptable, because symposia may embrace both types of presentation.

Contents

Preface

INTEREST IN THE EFFECTS OF RADIATION on polymeric materials is rapidly increasing. Changes in the physical or mechanical properties of a polymer can be induced by small amounts of radiation; for example, even a few scissions or cross-links per molecule can dramatically affect the strength or solubility of a polymer. Such changes determine whether a particular polymer will have an application in industry.

The irradiation of polymers is widespread in many industries. For example, microlithography is an essential process in the fabrication of integrated circuits that involves the modification of the solubility or volatility of thin polymer resist films by radiation. The sterilization by radiation of medical and pharmaceutical items, many of which are manufactured from polymeric materials, is increasing. This trend arises from both the convenience of the process and the concern about the toxicity of chemical sterilants. Information about the radiolysis products of natural and synthetic polymers used in the medical industry is required for the evaluation of the safety of the process.

UV and high-energy irradiation of polymer coatings on metals and other substrates has been developed for various industrial purposes. Irradiation causes the modification of polymer properties, and these special properties form the basis of major industries in heat-shrinkable films and tubings, cross-linked polymers, and graft copolymers. In the aerospace industry, advanced polymeric materials are used in many applications because of their high strength and low-weight performance. However, these materials are subject to degradation in space by UV and high-energy radiation, as well as bombardment by oxygen atoms.

A thorough understanding of the fundamentals of polymer chemistry and the effects of radiation on polymeric materials is therefore critical for the aerospace and other industries. Further technological advances in this field will require the interaction and collaboration between basic and applied researchers.

This book emphasizes the technological significance of the effects of radiation on polymers and draws attention to the major interactions between fundamental science and advanced technology. Although the field of polymer radiation chemistry is not exhaustively covered, a sampling of the ongoing basic and applied research in this area is presented. Review chapters have been included that cover fundamental radiation chemistry, spectroscopic methods, materials for microlithography, and radiation-durable materials.

The authors and editors are indebted to the National Science Foundation, the Department of Science, and the IBM Corporation for providing financial support to the U.S.–Australia Workshop on Radiation Effects on Polymeric Materials, from which this book was developed. Our sincerest thanks are extended to Robin Giroux and the production staff of the ACS Books Department for their efforts in publishing this volume.

ELSA REICHMANIS
AT&T Bell Laboratories
Murray Hill, NJ 07974

JAMES H. O'DONNELL
University of Queensland
St. Lucia, Brisbane 4067
Australia

August 24, 1988

Chapter 1

Radiation Chemistry of Polymers

James H. O'Donnell

Polymer and Radiation Group, Department of Chemistry, University
of Queensland, St. Lucia, Brisbane 4067, Australia

Changes in the properties of polymer materials caused
by absorption of high-energy radiation result from a
variety of chemical reactions subsequent to the
initial ionization and excitation. A number of
experimental procedures may be used to measure,
directly or indirectly, the radiation chemical yields
for these reactions. The chemical structure of the
polymer molecule is the main determinant of the
nature and extent of the radiation degradation, but
there are many other parameters which influence the
behaviour of any polymer material when subjected to
high-energy radiation.

Development of new applications of radiation modifications of the
properties of polymers in high technology industries such as
electronics and the exposure of polymer materials to radiation
environments as diverse as medical sterilization and the Van
Allen belts of space have resulted in a renewed interest in
fundamental radiation chemistry of polymers.
     The main features of the chemical aspects of radiation-
induced changes in polymers, which are responsible for changes in
their material properties are considered in this chapter.

TYPES OF RADIATION

High-energy radiation may be classified into photon and
particulate radiation. Gamma radiation is utilized for
fundamental studies and for low-dose rate irradiations with deep
penetration. Radioactive isotopes, particularly cobalt-60,
produced by neutron irradiation of naturally occurring cobalt-59
in a nuclear reactor, and caesium-137, which is a fission product
of uranium-235, are the main sources of gamma radiation. X-
radiation, of lower energy, is produced by electron bombardment
of suitable metal targets with electron beams, or in a

0097–6156/89/0381–0001$06.00/0
© 1989 American Chemical Society

synchrotron.   Photon radiation has a large half-distance for absorption compared to the range of particulate radiation.

Electron irradiation is normally obtained from electron accelerators to give beams with energies in the MeV range.   The corresponding penetration depths are then a few mm.   Much lower energy electron beams, e.g. 10-20 keV, are used in electron microscopy and in electron beam lithography.   A large proportion of the energy is then deposited in um thick polymer films.   The electron beams may be programmed to transfer a circuit pattern from a computer to a resist film of radiation-sensitive polymer.

Nuclear reactors are a source of high radiation fluxes. This comprises mainly neutrons and gamma rays, and large ionized particles (fission products) close to the fuel elements.   The neutrons largely produce protons in hydrocarbon polymers by "knock-on" reactions, so that the radiation chemistry of neutrons is similar to that of proton beams, which may alternatively be produced using positive-ion accelerators.

The effects of the radiation flux in space on polymer materials is now of considerable importance due to the increasing use of communications satellites.   Geosynchronous orbit corresponds to the second Van Allen belt of radiation, which comprises mainly electrons and protons of high energy.

Alpha particles cause intense ionization and excitation due to their large mass and consequently produce substantial surface effects.   Larger charged particles may be produced in positive ion accelerators.

ABSORPTION OF RADIATION

Photon radiation undergoes energy absorption by pair production (high energies, > 4 MeV), Compton scattering and the photoelectric effect (low energies, < 0.2 MeV).   In the photoelectric effect all of the energy of the incident photon is transferred to an electron ejected from the valence shell, whereas in Compton scattering there is also a scattered photon (of lower energy).   Thus, the radiation chemistry of photons occurs mainly through interaction of secondary electrons with the polymer molecules.

Electrostatic repulsion between high-energy electrons – produced from an accelerator, or by photon interaction with substrate atoms – and valency electrons in the polymer cause excitation and ionization.   The chemical reactions result from these species.

The absorption of high-energy radiation depends only on the electron density of the medium.   Mass density is a reasonable first approximation to electron density.   More accurately, and conveniently, the average value of the ration of Z/A for the atoms, where Z is the atomic number and A is the atomic mass, can be used to calculate relative dose.

The depth-profile of photon absorption is analogous to that for UV-visible light, i.e. I = Io exp(-Ad), where the mass energy absorption coefficient, u/g is used instead of the extinction coefficient.   Particulate energy absorption can be described by relative stopping powers.

The interaction of neutrons with organic molecules occurs mainly through knock-on of protons.   Thus, the radiation chemistry is similar to proton irradiation.   Radiation chemistry by positive ions is of increasing importance on account of ion implantation technology, plasma development and deposition processes, and cosmic irradiation.

UNITS.   Energy absorption has been traditionally expressed as dose rate in rad, corresponding to 10^{-2} J/kg.   The SI unit is the gray (Gy), which is 1 joule per kg.

Radiation chemical yields are conventionally expressed in G values for numbers of molecules changed or formed for 16 aJ (100 eV) of energy absorbed.

DEPTH PROFILE.   The secondary electrons produced by ionization processes from an incident beam of high-energy electrons are randomly directed in space.   Spatial "equilibrium" is achieved only after a minimum distance from the surface of a polymer in contact with a vacuum or gaseous environment (of much lower density).   Consequently, the absorbed radiation dose increases to a maximum at a distance from the surface (2 mm for 1 MeV electrons) which depends on the energy of the electrons.   The energy deposition then decreases towards zero at a limiting penetration depth.

TEMPERATURE RISE DURING IRRADIATION.   The chemical reactions which result from irradiation of polymers consume only a small fraction of the absorbed energy, which is mainly dissipated in the form of heat.   Thus, 0.1 MGy of energy absorbed in water will produce a temperature rise of 24 °C - and more in a polymer.

PRIMARY PROCESSES.   Absorption of high-energy radiation by polymers produces excitation and ionization and these excited and ionized species are the initial chemical reactants.   The ejected electron must lose energy until it reaches thermal energy. Geminate recombination with the parent cation radical may then occur and is more likely in substrates of low dielectric constant.   The resultant excited molecule may undergo homolytic or heterolytic bond scission.   Alternatively, the parent cation radical may undergo spontaneous decomposition, or ion-molecule reactions.   The initially ejected electron may be stabilized by interaction with polar groups, as a solvated species or as an anion radical.

$$P \xrightarrow{\text{\Large$\sim$}} P^{\ddot{+}} + e^-$$

$$\xrightarrow{\text{\Large$\sim$}} P^*$$

$$e^- \longrightarrow e^-_{th}$$

$$P^{\ddot{+}} + e^-_{th} \longrightarrow P$$

$$P^* \longrightarrow R_1{}^\cdot + R_2{}^\cdot$$

$$P^* \longrightarrow A^+ + B^-$$

$$P^{\ddot{+}} \longrightarrow C^+ + D^\cdot$$

$$P^+ + P \longrightarrow PX + E^+$$

$$e^- + S \longrightarrow e^-_{solv}, S^{\ddot{-}}$$

The radiation chemistry of polymers is therefore the chemistry of neutral, cation and anion radicals, cations and anions, and excited species.

SECONDARY REACTIONS. The reactions of the free radicals include (1) abstractions (of H atoms, with preference for tertiary H, and of halogen atoms), (2) addition to double bonds, which are very efficient scavengers for radicals, (3) decompositions to give both small molecule products, such as CO_2, and (4) chain scission and crosslinking of molecules.

$$R\cdot + R'H \longrightarrow RH + R'\cdot$$

$$R\cdot + R'Cl \longrightarrow R\,Cl + R'\cdot$$

$$R\cdot + CH_2 = CHR' \longrightarrow RCH_2 - \dot{C}HR$$

CAGE EFFECTS. When main-chain bond scission occurs in polymer molecules in the solid state to form two free radicals, the limited mobility of the resultant chain fragments must mitigate against permanent scission. This concept is supported by the increased yields of scission in amorphous compared with crystalline polymers. Similarly, the scission yields ar increased above, the glass transition, Tg, and melting, Tm, temperatures. There is also evidence from NMR studies of the changes in tacticity in poly(methyl methacrylate) that racemization occurs at a higher rate than permanent scission of the main chain, consistent with initial main-chain bond scission, rotation of the newly formed chain-end radical, and geminate recombination.

RADIATION-SENSITIVE GROUPS. Although the absorption of radiation energy is dependent only on the electron density of the substrate and therefore occurs spatially at random on a molecular scale, the subsequent chemical changes are not random. Some chemical bonds and groups are particularly sensitive to radiation-induced reactions. They include COOH, C-Hal, $-SO_2-$, NH_2, C=C. Spatial specificity of chemical reaction may result from intramolecular or intermolecular migration of energy or of reactive species - free radicals or ions.

Enhanced radiation sensitivity may be designed into polymer molecules by incorporation of radiation sensitive groups,and this is an important aspect of research in e beam lithography.

RADIATION-RESISTANT GROUPS. Aromatic groups have long been known to give significant radiation resistance to organic molecules. There was early work on the hydrogen yields from cyclohexane (G=5) and benzene (G=0.04) in the liquid phase, and of their mixtures, which showed a pronounced protective effect.

A substantial intramolecular protective effect by phenyl groups in polymers is shown by the low G values for H_2 and crosslinking in polystyrene (substituent phenyl) and in polyarylene sulfones (backbone phenyl), as well as many other aromatic polymers. The relative radiation resistance of different aromatic groups in polymers has not been extensively studied, but appears to be similar, except that biphenyl provides increased protection. Studies on various poly(amino acid)s indicate that the phenol group is particularly radiation resistant.

The combination of radiation-sensitive and radiation-resistant groups is interesting. Halogen substitution of the phenyl group in polystyrene results in high radiation sensitivity with inter-molecular crosslinking.

RADIATION-INDUCED CHEMICAL CHANGES IN POLYMERS

The molecular changes in polymers resulting from radiation-induced chemical reactions may be classified as:

1.   Chain crosslinking, causing increase in molecular weight. The continued crosslinking of molecules results in the formation of a macroscopic network and the polymer is no longer completely soluble, the soluble fraction decreasing with radiation dose.

2.   Chain scission, causing decrease in molecular weight. Many material properties of polymers are strongly dependent on molecular weight, and are substantially changed by chain scission. Strength - tensile and flexural - decreases, and rate of dissolution in solvent increases.

3.   Small molecule products, resulting from bond scission followed by abstraction or combination reactions, can give valuable information on the mechanism of the radiation

degradation. Gaseous products, such as CO_2, may be trapped in
the polymer, and this can lead to subsequent crazing and cracking
due to accumulated local stresses. Contamination of the
environment, e.g. by HCl liberated from poly(vinyl chloride), can
be a significant problem in electronic devices.

4. Structural changes in the polymer, which will accompany the
formation of small molecule products from the polymer, or may be
produced by other reactions, can cause significant changes to the
material properties. Development of colour, e.g. in
polyacrylonitrile by ladder formation, and in poly(vinyl
chloride) through conjugated unsaturation, is a common form of
degradation.

MECHANISMS OF SECONDARY REACTIONS. The primary processes
involved in absorption of radiation in polymers lead to the
expectation of free radical and ionic mechanisms for the
secondary chemical reactions. Electron spin resonance (ESR)
spectroscopy has proved extremely valuable for observation of
free radical reactions in polymers, where various radicals are
stabilized in the solid matrix at different temperatures.
Yields of radicals and kinetics of their transformations and
decays can be measured.
      Evidence for ionic reactions has been derived by the use of
specific scavengers (also applicable to radicals) and by
inference from ion-molecule reactions observed in the mass
spectrometer. Radical and ionic mechanisms can be written for
many chemical changes and the preferred pathway is likely to
depend on the irradiation conditions. e.g. temperature, and on
the presence of adventitious impurities, such as water, which
scavenge ions.

MEASUREMENT OF SCISSION AND CROSSLINKING

The changes in molecular weight may be used to determine yields
of scission and crosslinking. Average molecular weights may be
obtained by viscometry, osmometry, light scattering, gel
permeation chromatography and sedimentation equilibrium.
Equations have been derived which relate G(scission) and
G(crosslinking) to changes in Mn, Mw and Mz. Crosslinking
produces branched molecules and the relative hydrodynamic volume
(per mass unit) decreases compared with linear molecules.
Therefore, molecular weights derived from viscometry and gel
permeation chromatography will be subject to error.
      The equations relating Mn and Mw to radiation dose which are
most frequently used apply to all initial molecular weight
distributions for Mn, but only to the most probable distribution
(Mw/Mn = 2) for Mw. However, equations have been derived for
other initial distributions, especially for representation by the
Schulz-Zimm distribution equation.
      The use of Mz has been largely neglected in the evaluation
of crosslinking and scission in polymers, yet it is particularly
sensitive to higher molecular weight molecules produced by

crosslinking.   The equations for Mz currently available are difficult to use.   Mz can be obtained from sedimentation equilibrium experiments in the ultracentrifuge – an experimental procedure also largely neglected for synthetic polymers.

MOLECULAR WEIGHT DISTRIBUTIONS.   There is more information on scission and crosslinking available in the complete molecular weight distribution than in average molecular weights. Equations suitable for simulation of molecular weight distributions for any initial distribution and chosen values of G(scission) and G(crosslinking) have been developed and demonstrated.   The molecular weight distributions may be obtained by GPC (with the limitation of changes in relative hydrodynamic volumes) and by sedimentation velocity in the ultracentrifuge.

SOLUBLE/INSOLUBLE (GEL) FRACTION.   If crosslinking predominates over scission (when G(crosslink) > 4 G(scission)), the decrease in soluble fraction above the gel dose, may be used to derive G values for both processes.   An equation was derived by Charlesby and Pinner for the most probable molecular weight distribution and similar equations have been derived for other distributions.

Crosslinking yields can also be derived from the extent of swelling of the irradiated polymer (if the hydrodynamic interaction factor, X, between the polymer and the solvent is known accurately), or from stress relaxation measurements on elastomers.

NMR DETERMINATION OF SCISSION AND CROSSLINKING.   The methods described above using changes in molecular weight, soluble fraction or mechanical properties are related indirectly to the rates of scission and crosslinking.   They give no information about the nature of the crosslinks or the new chain ends.

For example, H crosslinks are considered to result from formation of a covalent bond between two different molecules. Two radical sites in close proximity may be produced by migration of the sites along chains or by formation of the sites in close proximity through H or X abstraction on the second molecule by an H or X atom formed by C-X scission on the first molecule.   H-links have been clearly demonstrated by ^{13}C NMR in irradiated polydienes.

NMR resonances attributable to Y-links have been reported in polyethylene after irradiation to low doses.   These crosslinks are suggested to be formed by reaction of a chain radical with a C=C double bond at the end of another molecule.

Methyl end groups resulting from main-chain scission in ethylene-propylene copolymers have observed by their characteristic 13C NMR resonance and determined quantitatively to give values of G(scission).

CLUSTERING OF CROSSLINKS.   The value of G(crosslink) obtained by quantitative ^{13}C NMR can be compared with values obtained by other methods, such as soluble fractions.   Much larger values have been obtained in radiation-crosslinked polybutadienes by NMR. This has been attributed to clustering of the crosslinks, so that the number would be under-estimated by swelling, solubility and mechanical property methods.

Clustering of crosslinks can be explained by a kinetic chain reaction occurring through the C=C double bonds. Crosslinking by the conventional vulcanization process with sulfur has been shown by NMR to proceed through the allylic hydrogen atoms. Thus, the mechanism of crosslinking is different in the two methods.

MORPHOLOGY EFFECTS

The rates of different chemical reactions in irradiated polymers are dependent on physical as well as chemical factors. Polymers may have crystalline and glassy or rubbery amorphous regions. The morphology may be quite complex with consideration of perfectness of crystallinity, orientation of molecules within amorphous regions and tie molecules.

There is well-established evidence that G values are usually greater in amorphous than crystalline regions, especially for crosslinking, which may not occur in crystalline regions, and greater in rubbery than glassy polymers. Increasing attention is being directed to the role of the interface between crystalline and amorphous regions. Radicals may migrate from the interior of the crystalline regions to these surfaces and be stabilized. Molecules at boundaries may be under stress and consequently exceptionally reactive.

LOSS OF CRYSTALLINITY. Radiation causes breakdown of the crystalline regions in polymers, although small increases may be observed at low doses, attributed to scission of tie molecules, reduction of molecular weight of polymer molecules in the amorphous regions and some secondary crystallization. These changes can be measured by thermal analysis techniques, such as differential scanning calorimetry (DSC). After large doses changes in the X-ray diffraction patterns become evident. The peaks broaden, due to decreasing crystallite size, and the ratio of crystalline peaks to amorphous halo decreases. Doses above 10 MGy cause substantial decreases in crystallinity.

TEMPERATURE EFFECTS

The rates of chemical reactions increase with temperature due to the greater proportion of molecules which have energies in excess of the activation energy and this will apply to radiation-induced secondary reactions in polymers. However, solid polymers are also characterized by their glass and melting transition temperatures. Substantial changes in molecular mobility occur across these transitions and the rates of chemical reactions are frequently greatly affected.

All chemical reactions are in principle reversible and this applies equally to polymerization. Therefore, formation of

active sites, particularly free radicals by chain scission, which
are identical to propagating radicals, can lead to depropagation.
The probability of depropagation will increase with temperature
and can have an important role in the radiation degradation of
polymers with low activation energies for propagation. Thus,
poly(alpha-methyl styrene) and poly(methyl methacrylate) show
increasing amounts of monomer formation during irradiation above
150 and 200 °C, respectively.

EFFECT OF DOSE RATE

The main effects of dose rate are due to an increase in
temperature of the polymer and depletion of oxygen (for
irradiation in air) at high dose rates.   It seems unlikely that
direct effects of dose rate should occur for electron, gamma and
X irradiation, due to the low spatial density of the ionizations
and excitations.

EFFECT OF STRESS

Research studies of radiation effects on polymer materials are
normally carried out on samples in powder, granule, film or sheet
form in a completely unstressed condition.
     There is evidence for both UV and high-energy irradiation of
polymers that simultaneous application of an applied stress with
irradiation can decrease the lifetime to failure.   This occurs
through an increase in the rate of creep, and is proportionately
greater at low stresses.   The cause of this stress-enhancement
of radiation-induced degradation is not understood.   Suggestions
have been made of localized heating or of increased
susceptibility of tie molecules, but further investigation of
this field is necessary, and important.
     Polymer materials are frequently used under stress loadings
and these may be concentrated at certain parts of the structure.
Thermal stresses may be induced by non-uniform heating or by
differential expansion coefficients; the latter may be an
important factor in the degradation of fibre-reinforced
composites in the radiation environment of space.
     Stresses may also be produced locally in polymers by
trapping of gaseous products during irradiation.   Processing of
polymers, as for example by injection moulding, film extrusion,
including uni- or bi-axial orientation, or solvent casting
without annealing, may also produce inbuilt stresses which
sensitize the polymer to degradation by radiation.

STRUCTURAL CHANGES IN POLYMERS

The chemical structures of polymers will be changed by the
evolution of small molecule products. The formation of C=C bonds
in the polymer backbone by loss of H_2 from hydrocarbon polymers,
or HCl from PVC, is well established and leads to colouration of
the polymer, especially with increasing sequence lengths of
conjugated unsaturation.   Carboxylic acid groups are

particularly susceptible to decomposition with liberation of CO
and CO_2. Short branches are preferentially lost from polymers
on account of the lower bond strength of attachment to the
backbone chain and the increased ability for diffusion of the
alkyl fragment from the scission site. NMR, UV and IR
spectroscopy have been used to observe these structural changes.

SMALL MOLECULE PRODUCTS. The small molecule products from
irradiation of polymers include hydrogen, alkanes and alkenes, CO
and CO_2, SO_2, H_2O and HCl depending on the chemical composition
of the polymer. They may be partly evolved and partly trapped
in the polymer according to their volatility, the sample
dimensions and the temperature.
    Gas chromatography can be used to determine the yields of
volatile products in very small amounts, especially by breaking
an ampoule of irradiated polymer in the injection system of the
chromatograph. The identity of the products, and their yields
can be determined with very high sensitivity by using mass
spectrometry, including the combination of GC and MS. Less
volatile products can be determined by liquid chromatography,
including HPLC, although this is a relatively unexploited area of
investigation.

LOW MOLECULAR WEIGHT MODEL COMPOUNDS. The mechanisms of
radiation effects on polymers are frequently investigated by
studies of low molecular weight model compounds. Analysis of
the chemical reactions is much easier than with high molecular
weight polymers. Thus, N-acetyl amino acids can be studied as
model compounds for poly(amino acid)s and hence for proteins.
    However, the chemical changes observed in low molecular
weight compounds can be quite misleading as models for polymers.
Difficulties include the high concentration of end groups, e.g.
COOH in N-acetyl amino acids, which can dominate the radiation
chemistry of the models. Low molecular weight compounds are
usually crystalline in the solid state and reactions such as
crosslinking may be inhibited or severely retarded.

ENVIRONMENT FOR IRRADIATION

Much research into radiation effects on polymers is done with
samples sealed under vacuum. However, polymer materials may, in
practical applications, be subjected to irradiation in air. The
effect of irradiation is usually substantially different in air,
with increased scission at the expense of crosslinking, and the
formation of peroxides and other oxygen-containing structures.
Diffusion rates control the access of oxygen to radicals produced
by the radiation, and at high dose rates, as in electron beams,
and with thick samples, the behaviour may be similar to
irradiation in vacuum. Surface changes may be quite different
from bulk due to the relative availability of oxygen.
    Irradiation of polymers in atmospheres of other gases offer
the possibility of a variety of chemical modifications of the

polymer molecules, especially at the surface.   This may enhance
scission or crosslinking, or alter the material properties.

ENERGY TRANSFER

Although the deposition of radiation energy is spatially random
on the molecular scale, the chemical changes are not random.
The selectivity of chemical change can be correlated with the
sensitivity of some chemical groups to radiation-induced
reactions and the resistance of others.   Transfer of the
absorbed energy to these reactive groups is necessary and there
is increasing evidence that it may occur by both inter- and
intra-molecular processes.   The chemical structure of the
polymer chain may provide pathways for energy transfer or energy
trapping.   Copolymers can be designed to control these
processes.

POLYMERIZATION OF OLIGOMERS

Crosslinking of oligomers (low molecular weight polymers) is
effectively a type of polymerization.   It is the basis of
radiation curing of surface coatings.   Radiation-sensitive
groups, such as double bonds in acrylates and methacrylates,
enhance polymerization and chain extension.   High dose rates are
used, mainly with electron irradiation, to achieve high
conversions in a few seconds and to take advantage of relatively
low diffusion rates to avoid oxygen inhibition.

LANGMUIR-BLODGETT FILMS

There is increasing interest in very thin polymer films for non-
linear optical effects in electronic applications.   Layers of
controlled orientation only a few molecules thick can be prepared
on glass or metal substrates by the Langmuir-Blodgett technique
using a surface film trough.   Low molecular weight monomer films
can be polymerized, or polymeric films modified by electron
irradiation.   It is likely that this area of radiation effects
on polymers will develop greatly in the future.

IRRADIATION OF COPOLYMERS

The range of properties of polymers can be greatly extended and
varied by copolymerization of two or more monomers.   The effects
of radiation on copolymers would be expected to show similarities
to the homopolymers, but major differences from linear
relationships are often experienced.   Aromatic groups in one
monomer frequently show an intramolecular protective effect so
that the influence of that monomer may be much greater than its
mole fraction.   The Tg of a copolymer is normally intermediate
between the homopolymers, except for block copolymers, and this
can cause a discontinuity in radiation degradation at a fixed
temperature.

Consideration of the relationship between the effects of radiation on homopolymers and copolymers raises the question of the variation from homopolymer behaviour with sequence length. Every copolymer has a distribution of sequence lengths for each comonomer. At what minimum sequence length does methyl methacrylate not show the high scission of PMMA? The future will probably see the development of processes for making polymers with controlled mini-block sequences to maximize a number of properties such as scission yield, adhesion, flexural strength, Tg..

IRRADIATION OF BLENDS

The properties of polymer materials can e greatly extended by blending two or more homopolymers together. Blends may be classified as compatible or incompatible - although this does depend on the dimensions being considered. Compatibility is influenced by the molecular weight of the homopolymers and is enhanced in practice by incorporation of block copolymers and other compatibilizers. The effects of radiation on blends depend on the degree of compatibility and the extent of inter-molecular interaction (physically and chemically) between the different types of homopolymers.

CROSSLINKED NETWORKS

Some polymer materials, particularly biomedical materials and step-growth polymers, comprise crosslinked networks. The effect of irradiation on networks, compared with linear polymers, will depend on whether scission or crosslinking predominates. Crosslinking will cause embrittlement at lower doses, whereas scission will lead progressively to breakdown of the network and formation of small, linear molecules. The rigidity of the network, i.e. whether in the glassy or rubbery state (networks are not normally crystalline), will affect the ease of crosslinking and scission.

POST-IRRADIATION EFFECTS

Most irradiated polymers show a continuing change in properties for a long period after irradiation. These post-irradiation effects may be attributed to (1) trapped radicals which react slowly with the polymer molecules and with oxygen which diffuses into the polymer (2) peroxides formed by irradiation in the presence of air or trapped within polymers irradiated in vacuum or an inert atmosphere) and slowly decompose with formation of reactive radicals, usually leading to scission, (3) trapped gases in glassy and crystalline polymers which cause localized stress concentrations.

The consequences of post-irradiation effects in polymer materials are progressive reduction in strength, cracking and embrittlement. Some reduction in these effects can be achieved by annealing of the trapped radicals, addition of appropriate scavengers, release of trapped gases, and control of the morphology of the polymer.

SUMMARY

There are a great number of parameters involved in determining how the properties of polymers are changed by high-energy radiation. Relationships between chemical structure and radiation sensitivity are modified by the morphology of the polymer and the irradiation conditions.

BIBLIOGRAPHY

1. O'Donnell, J.H; Sangster, D.F. Principles of Radiation Chemistry; Edward Arnold: London, 1970.
2. Swallow, A.J. Radiation Chemistry of Organic Compounds; International Series of Monographs on Radiation Effects in Materials, Vol.2; Charlesby, A., Ed.; Pergamon: Oxford, 1960.
3. Charlesby, A. Atomic Radiation and Polymers; International Series of Monographs on Radiation Effects in Materials, Vol.1; Charlesby, A., Ed.; Pergamon: Oxford, 1960.
4. Chapiro, A. Radiation Chemistry of Polymeric Systems; High Polymers, Vol.15; Interscience: New York, 1962.
5. Dole, M. Fundamental Processes and Theory; The Radiation Chemistry of Macromolecules, Vol.1; Dole, M., Ed.; Academic: New York, 1972.
6. Dole, M. Radiation Chemistry of Substituted Vinyl Polymers; The Radiation Chemistry of Macromolecules, Vol.2; Dole, M., Ed.; Academic: New York, 1973.
7. O'Donnell, J.H.; Rahman, N.P.; Smith, C.A.; Winzor, D.J. Macromolecules 1979, 12, 113.

RECEIVED October 3, 1988

Chapter 2

Early Events in High-Energy Irradiation of Polymers

David F. Sangster

Lucas Heights Research Laboratories, Division of Materials Science and Technology, Commonwealth Scientific and Industrial Research Organisation, Private Mail Bag 7, Menai, New South Wales 2234, Australia

The early events following the passage of a high energy particle through a polymer leading to the production of radical species are reviewed. An outline is given of some of the considerations developed in radiation chemical studies that could lead to a better understanding of polymeric systems. The importance of track structure, the migration of species, the role of oxygen, the study of model compounds and the use of pulse radiolysis techniques are discussed.

Expressed in its simplest form, high energy ionising radiation interacts with materials to produce ionisation and excitation (in almost equal amounts) and lattice defects. ($\underline{1}$) The resulting species can further react to give free radicals,

$$M \rightsquigarrow M^{+\bullet} + e \qquad (1)$$
$$\downarrow \qquad \longrightarrow \quad R\bullet + R\bullet' $$
$$M \rightsquigarrow M^* \qquad (2)$$

which can cause monomers to polymerise,

$$R\bullet + nM \longrightarrow RM_n\bullet \qquad (3)$$

polymers to crosslink and degrade,

$$P_n \xrightarrow{R\bullet} X \qquad (4)$$

$$P_n \xrightarrow{R\bullet} P_{n-m} + P_m \qquad (5)$$

and, in mixtures, monomers to graft to polymers

$$P_n + mA \xrightarrow{R\bullet} P_nA_m \qquad (6)$$

0097–6156/89/0381–0014$06.00/0
© 1989 American Chemical Society

The chemical nature and morphology of the material determines which of these reactions are predominant.

Most studies of the effects of ionising radiation on polymers and polymerising systems consider the reactions subsequent to the formation of free radicals. Using the variety of spectrometric and other techniques which are available, the identity of the radicals can be readily ascertained and their reactions can be followed. It is the purpose of this paper to explore the interface between the physics of radiation interaction with the polymer molecule, embedded as it is in its morphology, and the physical chemistry of the reactions of the resulting species which is followed by the organic chemistry of the radicals giving ultimately the observed crosslinking, scission and, if a monomer is present, initiation and grafting. These are all manifestations of the one set of phenomena following from the production of lesions by the early stochastic events. It is proposed to point up areas of research which could lead to a better understanding of what is happening in polymeric systems.

Interaction of Radiation with Matter

The fundamentals of the interaction of radiation with matter have been considered in several texts (1-8).

Electrons. A high energy electron (1 MeV ≡ 0.16 pJ) from an electron accelerator or beta-emitting radionuclide interacts with the orbital electrons of the material through it is passing. If the energy transfer in a single collision is large enough, ionisation will result giving an energetic electron and its positive geminate partner. (Equation 1)

A glancing collision will produce excitation of the atom or molecule (Equation 2) and the subsequent decay can result in luminescence or merely heat. A superexcited species may disintegrate giving ions, usually an electron and its positive geminate partner (Equation 2 → 1), or a pair of radicals. The original electron will continue along its path, perhaps with some deviation and with reduced energy (both energy and momentum will be conserved) undergoing further interactions.

The electrons produced during the first and subsequent collisions may be energetic enough to cause further ionisation and excitation (delta rays) until they reach the subionisation and subexcitation levels characteristic of the material. Finally, they undergo rotational-vibrational interactions until they reach thermal energies.

The statistical uncertainties in these processes are reflected in "straggling", a term which describes the variations in the number of collisions undergone by any one particle and the amount of energy transferred at each collision. There is a corresponding variation in the penetration or range of particles in a material. One MeV electrons have an average range of 4.3 mm in unit density material.

Following the ejection of an inner shell electron, the vacancy is filled from a higher energy shell, the energy released causing the emission of Auger electrons.

Since the secondary, tertiary, etc. electrons tend to have a nett velocity forward in the direction of the primary, the energy

deposited in the material will build up with depth until a
saturation or equilibrium value is reached and thereafter it will
decrease.

Energetic Photons. The gamma ray photon emitted during the
decomposition of an unstable radionuclide such as cobalt-60 (two
photons per disintegration (1.17 and 1.33 MeV)) can interact with an
orbital electron of the material by the process of Compton
interaction producing ionisation. The energies of the resulting
electrons from several such events constitute a continuum with the
average energy being approximately one half that of the reacting
photon (580 keV for cobalt-60 gamma radiation).

The degraded photon has the balance of the energy and undergoes
further interaction - mostly Compton interaction - until, at a
photon energy of about 50 keV for light material, the predominant
energy transfer process becomes the photoelectric effect whereby all
the energy of the photon is transferred to the resulting energetic
electron.

In the case of very high energy photons a third process,
production of an energetic electron/positron pair by photon/nucleus
reaction, becomes important. The threshold is 1.022 MeV but the
energy absorption exceeds that of Compton absorption only for
photons of energy greater than about 20 MeV for light materials, so
this effect can be neglected. Similarly, photonuclear reactions are
insignificant below 5 MeV.

Since approximately five or six interactions on the average are
required to convert cobalt-60 photon energy totally into electrons,
most of the events are produced by these secondary and subsequent
electrons. For all three of these processes then, gamma or
X-radiation acts effectively as if a number of tiny electron
accelerators were each producing a single energetic electron inside
the material. Thus the chemical reactions for electron beam and for
high energy photon radiation will be identical. The main difference
is that whereas external 1 MV electron beam radiation penetrates
only a few mm into a polymer surface, gamma radiation is absorbed
exponentially throughout the bulk of the material. Once again there
is a build up of events with depth and this is important in
estimating the dose to thin samples.

Ion Beams. Essentially, irradiation of materials with ion beams -
energetic protons or ions of higher mass, alpha particles, protons
produced by knock-on processes from fast neutrons, fission fragments
etc., also produces excitation and ionisation and secondary
energetic electrons by Coulombic interaction. The essential
difference between ion beam and electron or gamma radiation is in
the geometrical distribution of these events, which will be
considered under "Track Structure". Towards the end of their range
(22.5 μm for 1 MeV protons in unit density materials) there is a
sudden increase in energy deposition known as the "Bragg peak".

Ion beams can also cause nuclear reactions in the material but
we will not consider these further.

Lattice Displacements. In the above the question of lattice
displacements has not been covered. These will be apparent in hard

glassy or crystalline material particularly when subjected to ion bombardment. They are also observable in electron microscopy. They are rarely given much consideration in radiation chemistry.

Time Scale. At a time of 10^{-14} s after the passage of the high energy particle the super-excited molecules have relaxed or dissociated. At a time of 10^{-13} s the electron has been thermalised and some intra spur reactions have occurred. Spur reactions are complete by 10^{-8} to 10^{-7} s and the radiolytic species have begun to diffuse away into the bulk of the solution where they may react with dissolved solutes.

Comparison with Photochemistry. At a first approximation each electron of the material in the path of the ionising radiation or its secondary electrons has an equal probability of interaction. This has important consequences. Statistically a major component will interact in preference to a minor one – the proportion being controlled by their relative electron densities in the material. High energy ionising radiation is ubiquitous. Unlike light photons in photochemical reactions, it does not seek out specific bonds or structures. This lack of selectivity is a disadvantage in that it is not possible to favour certain reactions by using selected energies; but it also has advantages in being more universally applicable without having to add sensitising agents.

There are many subtleties in the interaction of ionising radiation with matter; for more details, reference should be made to the many texts (2-4), reports (5-8) and papers in the literature.

Track Structure.

The distribution of ionisation and excitation events throughout the material, i.e. the picture of the tracks, proposed by Mozumder and Magee (9-11) is now generally accepted. There are alternative treatments (12,13).

Spurs, Short Tracks and Blobs. For high energy electrons or gamma photons, the probability of interaction with the orbital electrons of the material is low so events are widely separated. These are called "isolated spurs". Each isolated spur will probably contain one to three ion pairs or excited species. It is not possible to be more definite but authors approaching the problem from different points of view usually arrive at answers within those limits. Some reaction between the species within an individual spur is possible but most probably the species will become separated and diffuse into the bulk of the solution. Attraction between the electron and its geminate positive partner leading to reaction between them will be considered in a later section. These tracks are often described as resembling "a string of beads". The analogy is valid provided that the beads are considered as being of different sizes, randomly separated by vast distances (average 200 nm decreasing as the energy falls), not joined together other than temporally, not quite linearly aligned, and having branch "strings" at 4 μm intervals on the average. The "branch tracks" are caused by high energy knocked-on electrons.

Electrons of energy less than about 5 keV have a much higher probability of interacting with the electrons of the material and events become increasingly close so that they begin to overlap forming cylindrical "short tracks". Chemical reactions between the radiolytic species are thus much more probable. Consequently, there is a fundamental chemical difference between these entities and isolated spurs although the distinction is not a sharp one.

Finally, towards the end of the electron track ($<$ $\approx$ 500 eV although more recently 1600 eV has been preferred (14)) there are densely ionised, pear-shaped regions called "blobs". Electrons of energy less than about 100 eV (being the upper limit of oscillator strength for light atoms in molecules) are no longer able to transfer energy effectively, other than to vibration levels.

The approximate distribution of absorbed energy between these three entities for liquid water have been calculated (10):

Electron Energy	Isolated Spurs	Short Tracks	Blobs
10 MeV	76	16	8
1 MeV	65	20	15
20 keV	38	50	12

Somewhat different proportions will apply for other materials but the picture will be similar. The energy limits are based on sound reasoning but include some intuition and there is a statistical distribution of events and distributions.

The picture has not been confirmed experimentally because time scales of less than 10 ps are not accessible at present and there are difficulties envisaged in reducing this limit below 1 ps. However, as a theoretical model it fits much of the experimental data and is of much greater value than that which uses the continuous slowing down approximation whereby energy is assumed to be deposited continuously along the track.

High energy protons behave in a similar way to electrons. For energies greater than one MeV the species exist predominantly as close isolated spurs arranged as a core containing about one half the energy produced by glancing collisions and a much less dense penumbra of delta tracks of high speed knock-on electrons which contain the other half of the absorbed or transferred energy. Radiolytic species are close enough for reaction with each other to compete with reaction with solutes. At lower energies the overlap and hence radical recombination becomes progressively more important. Corresponding to one MeV for protons the limits for high energy He nuclei (α particles) and C nuclei are approximately 10 MeV and 100 MeV, respectively. Thus there is no essential qualitative difference between the stubby tracks formed by heavy ions and the short tracks formed by intermediate energy electrons.

Geminate Recombination.   Coulombic forces attract the ionisation electron back to its geminate positive partner (reaction (1) $\rightarrow$ (2)).   In water and other polar media, these forces are modified by the high dielectric constant and the possibility of solvation, giving a high probability of escape. This results in a high yield of free electrons which can be measured by radiation chemistry techniques. On the other hand, in hydrocarbons and nonpolar media the Coulombic forces increase the probability of return and recombination to give an excited state and radical formation. The free electron yield is correspondingly low; further, it can be measured electrically. The thermalisation distance of the electron is typically 2 nm in polar liquids and 5-20 nm in aliphatic hydrocarbons.

Should there be a chemical group which reacts with the electron in either the molecule of the material or the solute, the geminate combination will be delayed and the electron yield may be correspondingly reduced (15-18).

Free electrons can be trapped in solid media in defect locations from which they can be released thermally or photolytically.

Intratrack Reactions. Earlier it was noted that chemical reactions could occur between the species before they could diffuse away from each other. In the case of irradiated water, for example, this will result in a greater yield of the molecular species, H_2 and H_2O_2, and a lower yield of the radical species, $\cdot OH$, e and $H\cdot$. Additionally, the amount of solute present in a short track or blob is limited and this may conceivably alter the course of reactions. In other respects, once the radiolytic species diffuse away from their place of origin into the bulk of the solution, they follow conventional solution behaviour. However, even in homogeneous solutions there will be differences attributable to the three track entities because of differences in the yields of molecular products formed by reactions between primary radicals and the yields of radical species which are able to escape out of the tracks. Thus it is possible to consider the overall yield of products as the sum of the individual yields from the three track entities, duly adjusted for the relative contributions from each, giving a weighted average equivalent to the observed experimental yields (10,19).

More sophisticated calculations (14,20), using either stochastic Monte Carlo or deterministic methods, are able to consider not only different irradiating particles but also reactant diffusion and variations in the concentration of dissolved solutes, giving the evolution of both transient and stable products as a function of time. The distribution of species within the tracks necessitates the use of nonhomogeneous kinetics (21,22) or of time dependent kinetics (23). The results agree quite well with experimental data.

Nonpolar Systems. Most of the early theoretical studies on radiation action were carried out on water and aqueous solutions. This was a consequence not only of its importance in radiation biology but also of the greater amount of experimental data and the simplicity of its radiation chemical reactions as compared with organic systems. Recently, however, more studies on non-polar systems such as alkanes have appeared (24). It is a long step to solid polymers but methods are being continually refined.

Absorbed Dose. In comparing different types of radiation and various chemical systems, the observed effect is usually expressed

in terms of the energy absorbed or dose since this physical quantity
can be measured and evaluated. Fundamentally however, the important
quantity is the number of collisions and their proximity or, better
still as far as the chemist is concerned, the number of each species
produced and their distributions followed by their movement and the
subsequent reactions. Of these, absorbed energy (one gray (Gy) =
one joule per kg) is readily quantified. The radiation chemical
yield is usually expressed as the G value - the number of molecules
changed per 100 eV or 16 aJ absorbed or the number of moles changed
per 9.648 MJ.

Linear Energy Transfer. The concept of linear energy transfer (LET)
has been a useful one in rationalising the variation in radiolytic
yields found for different types of radiation. It is still widely
used as it expresses in a single parameter some measure of the
"quality" of the radiation - the amount of energy deposited in a
given distance.

   Formally, the linear energy transfer is defined as

$$L_\Delta = (dE/dl)_\Delta$$

where dl is the distance traversed by the particle and dE is the
mean energy loss due to collisions with energy transfers less than
some specified value Δ. The units keV/μm are usually used. Track
average LET, dose average LET and LET distribution are related
quantities. High energy electrons or photons have low LET ($<$ 0.2
keV/μm) whereas ion beams or low energy ($<$ 1 keV) electrons have
high LET ($>$ 10 keV/μm). It is thus an indication of the proportion
of spur overlap. However, it does have grave limitations (5) in
bearing little direct relationship to the track structure as now
envisaged. In particular it takes no account of the size of the
site in which energy is deposited nor of the variations within the
various entities.

Microdosimetry. To explain some of the effects of radiation in
relation to a small susceptible target in the biological cell, Rossi
and his coworkers developed the concept and experimental techniques
to measure the microscopic distribution of energy deposited in
irradiated materials (8). The energy actually deposited can then be
related to the size of any microdomain.

   The lineal energy is defined as

$$y = \epsilon / \bar{l}$$

where ϵ is the energy imparted to the matter by a single event in a
volume of mean chord length $\bar{l}$ - which is 2/3 the sphere diameter.
The unit is J m^{-1} or keV/μm the same as for LET. The specific
energy (imparted) is

$$z = \epsilon / m$$

where m is the mass of the material in the sphere. The unit is the
gray, Gy (J kg^{-1}). ϵ can be measured in an ionisation chamber
or proportional counter attached to a multichannel analyser. The

gas of the counter replaces a small mass of the material. The lower limit of equivalent diameter is about 0.3 μm although the variance of doses has been evaluated experimentally in volumes as small as a few nanometres. A related parameter, the proximity function, is a measure of the probability that energy transfer points are separated by a given distance or contained within a given shell. Theoretical calculation based on known parameters, using either analytical or Monte Carlo methods, are well developed.

Applications to Polymers.

The relevance of these concepts to polymer studies will now be considered.

Polymers and Radiation Biology.   In systems having microregions highly sensitive to radiation,the amount of energy deposited or the number of events created in these regions could be critical. Examples of such systems are to be found in biology.   Charlesby (25-27) has remarked on the similarities between the effects of radiation in biological and in polymeric systems and on the possibility that studies in either field should contribute to our understanding of what is happening in the other.   For example, in radiation biology there has been for many years a theory of dual radiation action whereby two events must come together within a microsphere and within a critical time interval to score a kill (28).   The parallel with two radicals on neighbouring polymer strands bridging together to form a crosslink is apparent.

In polymers there are microregions undergoing reactions different from those occurring in the neighbouring material when exposed to radiation.   For example, the crosslinking behaviour in amorphous polyethylene differs from that in the crystalline phase (29).

Migration of excitation away from the site of the original radiation event has been invoked in polymers and has also been observed in biopolymers (30,31).

Radiation Quality.   In radiation biology, microdosimetry has generally been applied to studies involving low doses of radiation. Application to situations in polymer science requiring high doses in order to detect and quantify a response may not be straight forward since the later tracks can overlap existing lesions. However, using simulated tracks it would be possible to calculate the energy deposited in a thin film of polymer by various types of radiation. A thin bare film several microns thick will allow most of the delta rays to escape, thus giving a predominance of isolated spurs for electron irradiated samples and of overlapping tracks for ion beam irradiated samples.

Another source of overlapping tracks are the knock-on protons produced during fast neutron irradiation of a hydrogenous material.

Recent studies on LET effects in 2 mm thick samples of polymers have shown no difference between fast neutron and Co60 gamma irradiations for poly(olefins) (32), a small difference for poly(methylmethacrylate) (33) and a significant difference for poly(styrene) (34). This parallels the differences in yields found for aromatic compounds vis a vis those found for olefins (35). On the other hand the secondary carbon radicals formed by neutron irradiation of the model compound n-eicosane are less mobile than those formed by gamma irradiation (36).

A report (37) on the effect of different types of radiation on the elongation-at-break of certain commercial cable insulating materials pointed to several aspects requiring fundamental investigations on well-characterised materials under defined conditions. It showed the importance of anti-oxidant stabilisers, particularly in relation to long term ageing. It was concluded that the differences observed could be attributed to dose rate effects rather than to the types of radiation studied. The irradiated samples were standard dumb bell shaped tensile samples.

Migration of Excitation. Some of the more puzzling features of organic compound radiolysis have been explained by migration of excitation or of charged species (38). The migration of excitation along a polymer chain could rapidly dissipate the absorbed energy throughout the electron subsystem, rupturing the weaker bonds and those near embedded atoms and defects (39,40). Presumably the resulting radicals can not migrate as rapidly. The true situation is almost certainly more complex than that, as evidenced by the occurrence of morphology effects and post-irradiation events. Hydrogen atom hopping between chains, long distance tunnelling and intermolecular and intrachain transfer of excitation (41) have also been proposed. Since migration is slower at low temperatures thermoluminescence (42) can be expected to give a better insight. Perhaps time-resolved luminescence studies, which can access shorter times than can electron spin resonance techniques and measure something different, could assist here.

Recently, the time-dependence kinetics of such phenomena as luminescence has been interpreted in terms of detrapping of species (43). An alternative treatment involves the migration of an excited particle in a lattice (44).

Electrical Conductivity. Measurements of the electrical conductivity of materials under irradiation, particularly time-resolved conductivity following a pulse of electrons, can provide information on mobilities and trapping mechanisms, free electron yield, and thermalisation times and path lengths. Some measurements have been done on polymers (45-47) particularly because of their technological importance in the electronics industry (37). More fundamental work is needed.

Radiation Degradation. The radiation degradation of polymers which usually means main chain scission is of industrial importance in polymer utilisation. Radiation degradation is also made use of in electronic chip manufacture, in the commercial recovery of poly(tetrafluoroethylene) and in controlling the molecular weight distribution of such products as poly(ethyleneoxide) and acrylic paints. The radiation degradation of organic materials and components was the subject of an International Atomic Energy Agency consultants' meeting (48) and the degradation of thermoplastic polymers in air has been reviewed by Wilski (49).

The Role of Oxygen. The role of oxygen in irradiated polymer systems is a complex one. The surface of polymers irradiated in air can become oxidised and free radicals within the bulk can form peroxy radicals as oxygen subsequently diffuses in time (50). By irradiation in air, excess monomer can be removed; for example from polymer destined to be used for containing foods or beverages (51).

Oxygen can be implicated when main chain scission is involved and in surface oxidative degradation. Schnabel (52,53), in an extensive study of the degradation of polymers in solution, has shown that the presence of oxygen can (a) promote main chain degradation by forming peroxyl radicals which may prevent crosslinking; (b) inhibit main chain degradation in the case of polymers in which scission normally predominates; and (c) fix mainchain breaks in the case of polymers in which rapid repair of scission normally occurs. Which of these three actions is the dominant one depends on the polymer. Studies in solutions can help in the understanding of the possible reactions occuring in solid polymers which are more difficult to study experimentally.

Some studies on the effect of antioxidants have been reported (54-57). Atmospheric oxygen inhibits the polymerisation of coatings etc. as an industrial process. This still awaits a satisfactory alternative to using an inert atmosphere.

Model Compounds. Many of the complexities associated with practical polymers and the many simultaneous reaction pathways can be avoided by using model compounds. A typical one is eicosane ($n-C_{20}H_{42}$) which forms single crystals similar to those in the crystalline regions of polyethylene (58-60). More work is needed on similar compounds and on low molecular weight oligomers to establish both the species produced and the kinetics of their reactions (61,62).

Other Considerations. There are other approaches which could help in our understanding of the action of radiation.

Radiolysis v. Photolysis. The fundamental differences between the early interaction with materials of high energy ionising radiation and that of ultraviolet light have been pointed out above. Nevertheless, they can often lead to the production of the same radicals. Thus advances in one field can crossfertilise advances in the other. There are, however, few examples of the same polymer being studied under both types of radiation (63).

High Pressure. Irradiation under high pressure increases the crosslinking yield of some polymers (64-66). This has mechanistic implications which need to be further explored.

Pulse Radiolysis. Surprisingly few studies have been carried out using the powerful technique of pulse radiolysis to study radiation reactions in polymers although an increasing number of studies on organic systems (67) and on monomers (68,69) are appearing.

Pulse radiolysis, using as time-resolved detection methods optical absorption, luminescence, electrical conductivity or electron spin resonance can be expected to give information on the formation of transient or permanent radiation products and on their movement.

Macroradicals are readily formed in solution by dissociative electron capture

$$P_n Cl_m + e \longrightarrow P_n Cl_{m-1} \cdot + Cl' \qquad (7)$$

thus enabling their reactions to be studied.

Ion Beams. Several investigations have been made on the effects of ion beam irradiation on simple chemical systems. Polymers have also been irradiated with ion beams as an extension of studies on the ion implantation of other materials. To date, most of these studies have been concerned with gross effects, sometimes through to carbonisation (70) but here is a field which could have an industrial potential.

Industrial Importance. Radiation processes are being used increasingly in Australia and world wide to modify the surface and bulk properties of polymers. Radiation crosslinking has become important commercially; deliberate degradation less so. The use of radiation polymerisation to effect the rapid curing of coatings and printing inks and in the manufacture of magnetic media has increased greatly over the last few years. Radiation grafting of biocompatible surfaces and functional groups onto substrates is also being used increasingly. In addition there are several examples of the use of radiation in specialty areas. In all of these one does not need to understand what is happening in order to make use of an effect but it does help in keeping out of trouble. The increasing sophistication in radiation chemical investigations will translate into our interpretation of the events which occur when polymers are subjected to radiation.

Radiation as a controllable source of free radicals is finding its place in science and in industry.

Literature Cited

1.   O'Donnell, J.H.; Sangster, D.F. Principles of Radiation Chemistry; Edward Arnold (Publishers) Ltd: London, 1970.
2.   Farhataziz; Rodgers, M.A.J., Eds. Radiation Chemistry-Principles and Applications; VCH: New York, 1987.
3.   Kaplan, I.G.; Miterev, A.M. In Advances in Chemical Physics; vol. LXVIII; Prigogine, I; Rice, S.A., Eds.; Interscience: New York, 1987; pp.255-386.
4.   Baxendale, J.H.; Busi, F., Eds.; The Study of Fast Processes and Transient Species by Electron Pulse Radiolysis; D. Reidel: Dordrecht, 1982.
5.   Linear Energy Transfer; International Commission on Radiological Units and Measurements: Washington, D.C., 1970; Report No. 16.
6.   Radiation Quantities and Units; International Commission on Radiological Units and Measurements: Bethesda, 1980; Report No. 33.
7.   Radiation Dosimetry: Electron Beams with Energies between 1 and 50 MeV; International Commission on Radiological Units and Measurements: Bethesda, 1980; Report No. 35.

8. Microdosimetry; International Commission on Radiological
    Units and Measurements: Washington D.C., 1983; Report No. 36.
9. Mozumder, A.; Magee, J.L. Radiat. Res. 1966, 28, 203-214.
10. Mozumder, A.; Magee, J.L. Radiat. Res. 1966, 28, 215-231.
11. Mozumder, A.; Magee, J.L. J. Chem Phys. 1966, 45, 3332-3341.
12. Green, A.E.S.; Daryashankar, Schippnick, P.F.; Rio, D.E.;
    Schwartz, J.M. Radiat. Research 1985, 104, 1-14.
13. Pagnamenta, A.; Marshall, J.H. Radiat. Research 1986, 106,
    1-16.
14. Magee, J.L.; Chatterjee, A. J. Phys. Chem. 1978, 82, 2219-2226.
15. Yoshida, Y.; Tagawa, S.; Tabata, Y. Radiat. Phys. Chem. 1986,
    28, 201-205.
16. Muller, E.; Naumann, W. Radiat. Phys. Chem. 1987, 29, 127-129.
17. Plonka, A. Radiat. Phys. Chem. 1987, 30, 31-32.
18. Tachiya, M. Radiat. Phys. Chem. 1987, 30, 75-81.
19. Sangster, D.F. In Biophysical Aspects of Radiation Quality;
    International Atomic Energy Agency: Vienna, 1971; p. 481-496.
20. Turner, J.E.; Magee, J.L.; Wright, H.A.; Chatteree, A.; Hamm,
    R.N.; Ritchie, R.H. Radiat. Research 1983, 96, 437-449.
21. Freeman, G.R. In Annual Review of Physical Chemistry;
    Rabinovitch, B.S.; Schurr, J.M.; Strauss, H.L., Eds.; Annual
    Reviews Inc: Palo Alto, California, 1983; Vol. 34, p.463-492.
22. Freeman, G.R. Kinetics of Nonhomogeneous Processes; Wiley: New
    York, 1987.
23. Plonka, A. Radiat. Phys. Chem. 1987, 30, 31-32.
24. Bartcyzak, W.M.; Hummel, A. J. Chem. Phys. 1987, 87, 5222-5228.
25. Charlesby, A. Radiat. Phys. Chem. 1977, 9, 28.
26. Charlesby, A. Radiat. Phys. Chem. 1985, 25, 287-289.
27. Charlesby, A. In Reference 2, p. 451, 473.
28. Reference No. 8, p.60.
29. Luo. Y.; Wang, G.; Lu, Y.; Chen, N.; Jiang, B. Radiat. Phys.
    Chem., 1985, 25, 359-365.
30. Reference No. 8. p.56.
31. Adams, G.E. Radiat. Res. 1985, 104, S-32.
32. Seguchi, T.; Haryakawa, N.; Yoshida, K.; Tamura, N.; Katsumura,
    Y.; Tabata, Y. Radiat. Phys. Chem. 1985, 26, 221-225.
33. Egusa, S.; Ishigure, K.; Tabata, Y. Macromolecules 1979, 12,
    939-944.
34. Egusa, S.; Ishigure, K.; Tabata, Y. Macromolecules 1980, 13,
    171-176.
35. Ogawa, M.; Hirashima, S.; Ishigura, K.; Oshima, K. Radiat.
    Phys. Chem., 1980, 16, 15-19.
36. Tabata, M.; Sohma, J.; Yang, W.; Yokata, K.; Yamaoka, H.;
    Matsuryama, T. Radiat. Phys. Chem. 1987, 30, 147-149.
37. Hanishch, F.; Maier, P.; Okada, S.; Schonbacher, H. Radiat.
    Phys. Chem. 1987, 30, 1-9.
38. Hummel, A.; Luthjens, L.H. J. Radioanal. Nucl. Chem. 1986, 101,
    293-297.
39. Kaplan, I.G.; Miterev, High En. Chem. 1985, 19, 168.
40. Zhadanov, G.S.; Smolyanskii, A.S.; Milinchuk, V.K. High En.
    Chem. 1986, 20, 27-31.
41. Kashiwabara, H. Radiat. Phys. Chem. 1986, 27, 361-362.
42. Fleming, R.J.; Hagekyriakou, J. Radiat. Prot. Dosim. 1984, 8,
    99-116.
43. Plonka, A. Radiat. Phys. Chem. 1986, 28, 429-432.

44. Dissado, L.A. Chem. Phys. Lett. 1986, 124, 206–210.
45. Tyutnev, A.P.; Abramau, N.V.; Dubenskov, P.I.; Saenko, V.S.; Vannikov, A.V.; Pozhidaeu, E.E. High En. Chem. 1986, 20, 509–514.
46. Tyutnev, A.P.; Abramau, N.V.; Dubenskov, P.I.; Saenko, V.S.; Pozhidaeu, E.; Vannikov, A.V. High En. Chem. 1987, 21, 148–153.
47. Dubonekov, P.I.; Tryutner, A.P.; Saenko, V.S.; Vannikov, A.V. High En. Chem. 1985, 19, 92–97.
48. IAEA Consultants' Meeting on Radiation Degradation of Organic Materials and Components, Cadarache France 1983 International Atomic Energy Agency, Vienna.
49. Wilski, H. Radiat. Phys. Chem. 1987, 29, 1–14.
50. Hikmet, R.; Keller, A. Radiat. Phys. Chem. 1987, 29, 15–19.
51. Salame, M.; Steingiser, S. U.S. Patent 4, 174, 043 (Nov 13, 1979).
52. Schnabel, W. Radiat. Phys. Chem. 1986, 28, 303–313.
53. Schnabel, W. J. Radioanal. Nucl. Chem. 1986, 101, 413–432.
54. Babic, D.; Gal, O.; Stannett, V.T. Radiat. Phys. Chem. 1985, 25, 343–347.
55. Gal, O.; Kostoski, D.; Babic, D.; Stannett, V.T. Radiat. Phys. Chem. 1986, 28, 259–267.
56. Cernoch, P.; Vokal, A.; Ciperova, D. Radiat. Phys. Chem. 1986, 28, 501–504.
57. Brede, O.; Hermann, R.; Mehnert, R. Radiat. Phys. Chem. 1986, 28, 507–510.
58. Hamanoue, K.; Kamantauskas, V.; Tabata, Y.; Silverman, J. J. Chem. Phys. 1974, 61, 3439–3444.
59. Chappas, W.J.; Silverman, J. Radiat. Phys. Chem. 1980, 16, 437–443.
60. Seguchi, T.; Haryakawa, N.; Tamura, N.; Tabata, Y.; Katsumura, Y.; Haryashi, N. Radiat. Phys. Chem. 1985, 25, 399–409.
61. Kroh, J. Radiat. Phys. Chem. 1986, 28, 415–424.
62. Dole, M. J. Phys. Chem. 1987, 91, 3117–3119.
63. Guillet, J.E. Photophysics and Photochemistry of Polymers; Cambridge Univ. Pr.: Cambridge, 1985; p354.
64. Burillo, G.; Ogawa, T. Radiat. Phys. Chem. 1985, 25, 383–388.
65. Kiryukhin, V.P.; Klinshpont, E.R.; Milinchuk, V.K. High En. Chem. 1985, 19, 86–91.
66. Milinchuk, V.K.; Klinshpont, E.R.; Kiryukhin, V.P. Radiat. Phys. Chem. 1986, 28, 331–334.
67. Mayer, J.; Szadkowska – Nicze, M.; Kroh, J. J. Radioanal. Nucl. Chem. 1986, 101, 359–367.
68. Kadhum, A.A.H.; Langan, J.R.; Salmon, G.A.; Edwards, P.P. J. Radioanal. Nucl. Chem. 1986, 101, 319–327.
69. Yoshida, H.; Ogasawara, M. J. Radioanal. Nucl. Chem. 1986, 101, 339–348.
70. Symposium Papers. Rad. Effects 1986, 98, 115–188.

RECEIVED August 10, 1988

Chapter 3

Spectroscopic Methods in Polymer Studies

Kenneth P. Ghiggino

Department of Physical Chemistry, University of Melbourne, Parkville 3052, Australia

The physico-chemical changes induced in polymers following exposure to radiation can be studied by a range of spectroscopic techniques. Recent developments in instrumentation and data analysis procedures in electronic, vibrational and magnetic resonance spectroscopies have provided considerable new insights into polymer structure and behaviour. The application of these spectroscopic methods in polymer studies are reviewed with emphasis on their utility in investigations of radiation effects on macromolecules.

Spectroscopic methods are now widely used in the polymer field as an analytical tool to probe structure and to obtain information on physico-chemical changes occurring in polymers and polymer additives. Spectroscopy utilizes the interaction of radiation with matter to provide details of molecular energy levels, energy state lifetimes and transition probabilities. This information in turn may be applied in studying chemical structure, molecular environment, polymer tacticity and conformation, and to monitor changes in these properties following external perturbations (e.g., mechanical stress, thermal treatment, radiation exposure). The advantages of spectroscopic measurements over other means of polymer characterization are that they are a non-destructive and rapid means of providing information at a molecular level. Exposure of polymers to ultraviolet and higher energy radiation can lead to extensive physical and chemical modification of polymeric materials. These changes in properties may have both detrimental and beneficial consequences in determining the end-uses of the polymer. Spectroscopy can provide a detailed insight into the mechanisms of polymer modification occurring under irradiation thus enabling control of the final material properties.

0097–6156/89/0381–0027$06.00/0
© 1989 American Chemical Society

There are several well written texts describing the theory and application of spectroscopy to polymer systems (1-3) to which the reader is referred for detailed information. Within the confines of this chapter, some aspects relevant to radiation effects in polymeric materials are reviewed with particular emphasis on new developments in instrumentation. The various spectroscopic methods can be distinguished by the energy of the transitions investigated (c.f. Figure 1) and this notation is employed in the sub-headings discussed below. Surface analysis techniques will not be described in detail in this article although mention may be made of ESCA (Electron Spectroscopy for Chemical Analysis) as one recent technique which may be used to analyze surface properties. ESCA monitors the kinetic energy of electrons detached from the sample following irradiation with an X-ray source. The detached electrons arise from atoms and molecules at the surface of the polymer (~2 nm depth) and thus information concerning chemical modifications (e.g., photo-oxidation) at surfaces can be obtained (1).

UV-Visible Absorption and Emission Spectroscopy

The absorption and emission of radiation in the near ultraviolet (UV) and visible regions of the electromagnetic spectrum are associated with electronic (and associated vibronic) transitions involving π- and/or n-electron systems of molecules. Synthetic and natural polymers absorb in the UV region and particularly strong absorption spectra are recorded for polymers containing aromatic and heteroaromatic groups (e.g., poly(styrenes), poly(vinyl naphthalenes), poly(vinyl carbazoles)). Polymers with chromophores exhibiting $n\pi^*$ transitions (e.g., C=O) exhibit weaker UV-absorption and these groups together with unsaturated carbon-carbon bonds which develop during radiation damage can be detected by electronic absorption spectroscopy.

The absorption and emission processes occurring in organic molecules including polymers can be discussed qualitatively with reference to the state diagram of Jablonskii (Figure 2). Absorption of radiation by the molecule leads to promotion of an electron from the ground singlet state (S_0) to a higher electronic state which, through conservation of spin, will also be a singlet state S_n. The energies of radiation required to promote electrons to higher energy states and the range of associated vibrational energy levels, results in the recorded absorption spectrum with the intensity distribution of the spectrum reflecting the relative probabilities of the transitions. Associated with each excited singlet state there is an electronic state of lower energy in which the electron spins are parallel, the triplet state. It should be noted that absorption spectra arising from any excited state of the molecule (e.g., S_1 - S_n absorption, T_1 - T_n absorption) or other transient

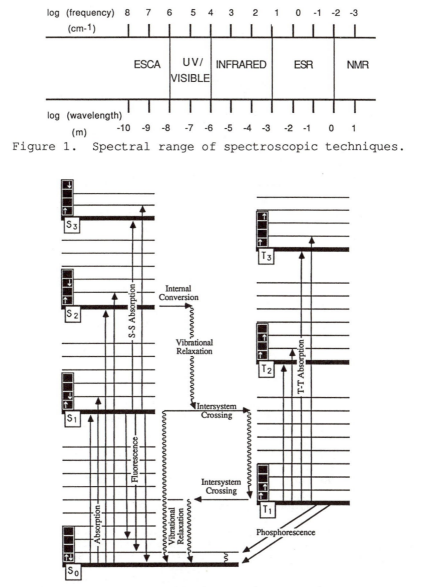

Figure 1. Spectral range of spectroscopic techniques.

Figure 2. Jablonskii diagram illustrating
photophysical processes in a polyatomic molecule.

species (e.g., free radicals, radical ions) can also be
recorded if the state can be sufficiently populated and
these experiments form the basis of the flash photolysis
technique.

The possible fate of excitation energy residing in
molecules is also shown in Figure 2. The relaxation of
the electron to the initial ground state and accompanying
emission of radiation results in the fluorescence spectrum
$(S_1 - S_0)$ or phosphorescence spectrum $(T_1 - S_0)$. In
addition to the radiative processes, non-radiative
photophysical and photochemical processes can also occur.
Internal conversion and intersystem crossing are the
non-radiative photophysical processes between electronic
states of the same spin multiplicity and different spin
multiplicities respectively.

The lifetime of the excited state will be influenced
by the relative magnitudes of these non-radiative
processes and thus time-resolved spectroscopy can provide
information on the dynamics of excited state depletion
mechanisms, e.g.,

		Process	Rate
$S_0 + h\nu$	$\rightarrow S_1$	Absorption	I_a
S_1	$\rightarrow S_0 + h\nu_F$	Fluorescence	k_F
S_1	$\rightarrow S_0$	Internal Conversion	k_{IC}
S_1	$\rightarrow T_1$	Intersystem Crossing	k_{ISC}
S_1	$\rightarrow$ Products	Reaction	k_R

$$\tau_F = (k_F + k_{IC} + k_{ISC} + k_R)^{-1}$$

The lifetime of the singlet excited state (the
fluorescence lifetime τ_F) is of the order of picoseconds
to 100 nanoseconds (10^{-12} - 10^{-7} seconds) and can now be
measured accurately using pulsed laser excitation methods
and other techniques. Since the radiative transition from
the lowest triplet state to the ground state is formally
forbidden by selection rules, the phosphorescence
lifetimes can be longer, of the order of seconds.

The UV-visible absorption and emission spectra and
excited state lifetimes of polymers are sensitive to
chemical structure, polymer conformation and molecular
environment and thus information concerning these
properties is accessible by electronic spectroscopy
measurements (4-6). One example of the application of
such measurements is given in Figure 3 which illustrates
the possible energy dissipation pathways which can occur
in a polymer containing aromatic side groups following
absorption of radiation.

Light initially absorbed by one chromophore on the polymer chain may very rapidly be transferred to a neighbouring chromophore by the process of non-radiative energy transfer (4-6). In this way, excitation energy can migrate along the polymer chain until it is trapped at sites of lower energy which subsequently undergo further photophysical/photochemical processes. The low energy sites may be excimers (excited state dimer complexes(5)) or species with lower excited state energies which have been deliberately incorporated in the polymer.

Since the emission from single excited chromophores, excimer sites and acceptor groups are spectrally distinguishable, fluorescence techniques can be used to characterize the excited state species present in the polymer. In Figure 4 the fluorescence spectra recorded from dilute solutions of poly(styrene) (PS) and a 1:1 alternating copolymer of styrene and maleic anhydride are shown. For PS, fluorescence from the excited aromatic phenyl chromophores is observed (fluorescence maximum 290 nm) together with a broad emission from excimer sites in the polymer (fluorescence maximum 330 nm). However in the alternating copolymer no excimer fluorescence is detected demonstrating that under these conditions excimer energy trap sites in PS must form between adjacent phenyl group chromophores along the polymer chain.

Recent developments in laser technology and fast detection methods now allow the kinetic behaviour of the excited state species arising from absorption of radiation by polymers to be studied on time-scales down to the picosecond region (5). An example of a time-resolved fluorescence spectrometer which can be used to study such ultrafast phenomena is illustrated in Figure 5 (7).

The commercially available laser source is a mode-locked argon-ion laser synchronously pumping a cavity-dumped dye laser. This laser system produces tunable light pulses, each pulse with a time duration of about 10 picoseconds, and with pulse repetition rates up to 80 million laser pulses/second. The laser pulses are used to excite the sample under study and the resulting sample fluorescence is spectrally dispersed through a mono-chromator and detected by a fast photomultiplier tube (or in some cases a streak camera (5)).

In the time-correlated single photon counting technique depicted in Figure 5, electronic pulses synchronised with the laser pulses are used to initiate a voltage-time ramp in the time-to-amplitude converter (TAC) while electronic pulses arising from fluorescence photons incident on the photomultiplier tube terminate the voltage ramp. The amplitude of the voltage pulse from the TAC, which is stored in a memory of a multichannel analyser operating in pulse height analysis mode, will be directly related to the time between excitation of the sample by the laser pulse and the detection of a fluorescence photon. Collection of many such events at a fixed emission wavelength results in a fluorescence decay curve which may

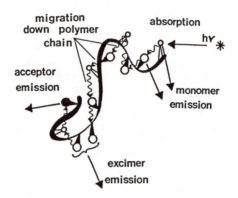

Figure 3. Energy relaxation pathways in a polymer containing aromatic side groups following absorption of light. (Reproduced with permission from Ref. 21. Copyright 1987 Chemistry in Australia.)

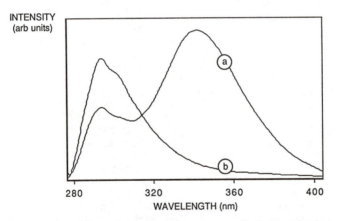

Figure 4. Fluorescence spectra of (a) poly(styrene) and (b) styrene/maleic anhydride alternating copolymer in tetrahydrofuran at 20°C. Excitation wavelength: 265 nm.

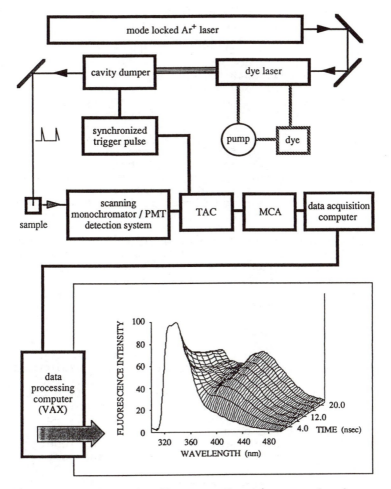

Figure 5. Schematic diagram of a time-resolved fluorescence spectrometer using a picosecond laser as an excitation source. Inset diagram: intensity/time/ wavelength surface for poly(acenaphthalene) in benzene at 20°C. Excitation wavelength 295 nm. (Reproduced with permission from Ref. 21. Copyright 1987 Chemistry in Australia.)

be analyzed to extract the fluorescence decay parameters
(5). Alternatively accumulation of fluorescence events at
a number of emission wavelengths and at various times
after excitation can produce three-dimensional hypersurfaces displaying the spectral, temporal and intensity
information simultaneously (Figure 5).
    The inset diagram in Figure 5 depicts the kinetic
behaviour of fluorescence from single excited chromophores
(fluorescence maximum 340 nm) and excimer sites
(fluorescence maximum 400 nm) following excitation of a
solution of poly(acenaphthalene) with laser pulses at 295
nm. Such hypersurfaces provide an overview of the excited
state processes occurring in polymers following
irradiation and can show where the energy initially
absorbed by the polymer finally resides and the rates of
various energy dissipation pathways. Absorption,
fluorescence and phosphorescence measurements have been
applied to the study of radiation effects on a wide range
of polymers and polymer additives (1-7). Chemiluminescence
is a further example of the application of electronic
spectroscopy measurements to polymer degradation studies.
The weak visible radiation emitted from polymers undergoing thermo-oxidative degradation has been attributed to
emission from electronically excited oxidation products
and thus chemiluminescence provides a means of detecting
and monitoring the incipient stages of certain polymer
decomposition processes.

Infrared and Raman Spectroscopy

A quantum description of the stretching and bending
vibrations of molecular bonds results in the assignment of
vibrational energy levels associated with each electronic
state of functional groups in polymers (1). Since bond
force constants and atomic masses determine the
vibrational frequencies, the energy levels will be
characteristic of the chemical groups present in the
polymer. Transitions between these vibrational energy
levels can be investigated using infrared and Raman
spectroscopies.
    Infrared (IR) spectroscopy has been widely used in
polymer studies for the assignment of molecular structure
and for monitoring changes in the arrangement of chemical
bonds (1,8-11).
Information concerning conformation, tacticity and
crystallinity may also be obtained (1). Vibrational
transitions accesssible to IR spectroscopy are governed by
the selection rule that there must be a change in dipole
moment during excitation of the polymer vibrations. Thus
symmetric vibrations which are detected by Raman spectroscopy are inaccessible to IR absorption measurements.
Polar groups, such as carbonyl (C=O) and hydroxyl (OH),
have a strong ground state dipole moment and show strong
IR absorptions at characteristic frequencies. The IR
spectrum can thus be used as a 'fingerprint' of molecular

structure and, since the positions of vibrational frequencies are sensitive to neighbouring chemical groups, conclusions concerning local environment can be made. A considerable volume of polymer IR spectra have been recorded and published (12).

Both conventional dispersive element IR spectrometers and Fourier Transform infrared (FTIR) interferometer based instruments are available to record IR spectra of polymers. In the conventional instruments a dispersive element, such as a grating or prism is used to measure the frequencies at which infrared radiation is absorbed by the sample. An interferometer (Figure 6) constitutes the basis of the FTIR instrument. In the interferometer the IR radiation is split into two paths and, after reflection from mirrors (one movable), the beams are recombined at the beam splitter. When the path lengths followed by the two beams are identical all wavelengths of radiation incident on the beam splitter add coherently and result in the maximum flux at the detector. At other positions of the movable mirror destructive interference of each wavelength at the beam splitter occurs and the flux at the detector will decrease. The interferogram (F(x)) produced by recording the radiation flux as the mirror undergoes translational movement has the form of a damped oscillation corresponding to:

$$F(x) = \int A(v) \cos(2\pi v) \, dv$$

where A(v) is the spectral intensity distribution containing the spectroscopic information. The spectroscopic data is extracted by a Fourier transform of the interferogram (11). The instrument is single beam and a blank must also be measured and subsequently subtracted. A computer carries out the Fourier transform calculations, performs various control functions and manipulates the data for display and interpretation.

FTIR instruments offer advantages in speed and higher signal-to-noise ratios compared to dispersive IR spectrometers. These advantages combined with the facility for extensive data processing have seen the FTIR technique find increasing applications in polymer studies (11).

IR spectroscopy has proved most useful in studying chemical modifications of polymers induced by external factors including radiation damage (1,13). Oxidation is an important degradation pathway following exposure to both heat and radiation. The strongly polar functional groups which are the products of oxidative damage (e.g., OH, C=O) are readily detected by IR spectroscopy and thus this technique can be used to follow the early stages of degradation. Attenuated total reflection (ATR) of IR radiation (1) may also be used to monitor surface modifications during degradation.

An example of the application of IR spectroscopy is in the photooxidation of poly(propylene) (1,3). During the early stages of oxidation absorption due to aldehydes

(1735 cm^{-1}) and ketones (1720 cm^{-1}) are apparent while at
latter times carboxylic acids (1710 cm^{-1}) can be detected.
In poly(ethylene) hydroperoxides (3550 cm^{-1}) are observed
during early stages of irradiation while FTIR has revealed
an increase in vinyl end groups, carbonyls and trans-
vinylidene double bonds (11). Correlations have been noted
between physical changes in the polymer and chain scission
processes detected by IR spectroscopy during photodegradation of poly(ethylene) (13).

The advent of lasers has assisted in the development
of Raman spectroscopy as a means of recording vibrational
spectra of polymers and other molecular systems. Raman
spectroscopy is based on inelastic light scattering and
uses monochromatic radiation in the visible region as the
excitation source. Analysis of radiation scattered from a
sample of molecules indicates the presence of frequencies
which are spectrally shifted to lower energies (Stokes
lines) and higher energies (anti-Stokes lines) compared to
the incident radiation. The spectrally shifted lines
arise due to the transfer of vibrational quanta between
the interacting radiation and the medium. The observed
transitions are governed by the selection rule that there
is a change in polarizability during the molecular
vibration and thus IR-inactive totally symmetric
vibrations may be observed. In order to discriminate the
Raman spectral lines from the strong Tyndall scattering of
the incident radiation, highly monochromatic radiation
available from laser sources (typically argon-ion or
krypton lasers) is preferred and double or triple mono-
chromators are often required to achieve the necessary
spectral resolution. Resonance Raman scattering may also
be observed if the frequency of the exciting radiation
corresponds closely to an electronic absorption bond. In
this case, the Raman lines arising from coupling of the
vibrations with the electronic transition are much
stronger than ordinary Raman scattering.

Further details of the theory and application of
Raman spectroscopy in polymer studies can be found
elsewhere (1,9). However, vibrational frequencies of
functional groups in polymers can be characterized from
the spacing of the Raman lines and thus information
complementary to IR absorption spectroscopy can be
obtained. In addition, since visible radiation is used
the technique can be applied to aqueous media in contrast
to IR spectroscopy, allowing studies of synthetic
polyelectrolytes and biopolymers to be undertaken.
Conformation and crystallinity of polymers have also been
shown to influence the Raman spectra (1) while the
possibility of studying scattering from small sample
volumes in the focussed laser beam (~100 μm diameter) can
provide information on localized changes in chemical
structure.

One new technique of potential importance to the
study of the interaction of radiation with polymers is
time-resolved Raman spectroscopy (14,15). In these

experiments excited states and transient species are
produced during excitation of the sample by a short laser
pulse. The Raman scattering induced by a second probe
pulse incident on the sample after a fixed time delay can
be used to characterize the vibrational frequencies and
hence the structure of the intermediate species. The
application of mode-locked tunable dye lasers has allowed
the technique to be extended down to the picosecond time
region. Transient species produced following absorption
of picosecond light pulses by bacteriorhodopsin (a protein
complex containing the retinal chromophore in the purple
membrane of <u>Halobacterium</u> <u>Halobium</u>) have recently been
undertaken (G.Atkinson, University of Arizona, personal
communication,1987) although the application of such
measurements to synthetic macromolecular systems has yet
to be fully investigated.

<u>Magnetic Resonance Spectroscopy</u>

Electrons and nuclei have a magnetic moment associated
with angular momenta of the particles. In the presence of
a magnetic field the degeneracy of discrete energy levels
associated with the magnetic moment is removed and
absorption and emission of radiation between these energy
levels may be observed (<u>1,16</u>), The energy difference
between the quantized levels depends on the magnetic field
strength and is given by:

$$\Delta E = h\nu = g\mu B$$

where g is a constant (g_e = 2 for free electrons; g_N = 0.1
to 6 for many nuclei, e.g., g_N = 5.5854 for a proton), μ
is the Bohr magneton (μ_B) for electrons and the nuclear
magneton (μ_N) for nuclei and B is the magnetic field flux
density.
        The technique of studying the absorption of
radiation by unpaired electrons in a magnetic field is
called Electron Spin-Resonance (ESR) spectroscopy while
the study of the resonance frequencies for nuclei is
classified as Nuclear Magnetic Resonance (NMR)
spectroscopy. Under external magnetic field strengths of
about 1 Tesla, ESR spectroscopy requires energies in the
microwave region (~10 GHz) to initiate transitions while,
since the interaction between nuclei and the magnetic
field is much weaker, the NMR technique uses lower energy
radio waves in the 1 to 5 metre band.
        Experimentally the sample is placed in a strong
magnetic field and, rather than the frequency being
scanned at a constant field strength to detect absorption
of radiation, in practice the frequency of exciting
radiation is kept constant and the magnetic field flux is
varied. Both ESR and NMR spectroscopy have found
widespread application in polymer studies and several
excellent texts describing the techniques are available
(<u>1,17-19</u>).

ESR spectroscopy requires the presence of unpaired electrons in the sample and thus it finds application in the study of triplet excited states, neutral free radicals and radical ions which may be formed in polymers following exposure to radiation. The degenerate energy levels of unpaired electrons are split into two in the presence of a magnetic field and a single resonance absorption might be expected. However, there are often interactions between the magnetic moments of other neighbouring nuclei (e.g., protons) and the electron leading to hyperfine splittings in the absorption spectrum (17). In this case, the electron and nuclear spins interact and impose slightly different energy levels on the original energy splitting arising from the effect of the magnetic field on the electron. Thus different resonant frequencies will be observed and the number and intensity of the absorption bands in the spectrum can provide information about the chemical environment of the unpaired electron (1,17).

For example, poly(methyl methacrylate) exposed to high energy or UV radiation gives a nine line ESR spectrum, as depicted in Figure 7. Analysis of this spectrum has indicated that the likely structure of the free radical responsible is (13):

$$\sim CH_2 - C^{\bullet}(CH_3)\,COOCH_3.$$

In poly(olefines) the metastable allyl radical is often observed by ESR techniques following irradiation:

$$\sim CH_2 - CH = CH - HC^{\bullet} - CH_2.$$

The capability to detect such species by ESR spectroscopy provides a means to analyse the mechanisms of polymer breakdown under irradiation (17,19). In addition, certain compounds used to photostabilize polymers against UV radiation act by scavenging the reactive radicals to form more stable radical species (e.g., hindered phenoxy radicals) and thus the performance of these stabilizers can be assessed by ESR methods (17).

The species present in polymers that can be studied by ESR are often highly reactive, short-lived and are present in low concentrations. However, developments in instrumentation have offered improvements in sensitivity and, combined with more reliable interpretation of data (1), the increasing application of this method of polymer characterization in studying radiation effects on polymers can be expected.

In contrast to ESR spectroscopy, which can only be used to study species with unpaired electrons, NMR spectroscopy is applicable to the investigation of all polymer samples. Nuclei with non-zero total nuclear spin (e.g., ^1H, ^{13}C, ^{19}F, ^{14}N) will have a magnetic moment which will interact with an external magnetic field resulting in quantized energy levels. Transitions between these energy levels form the basis of NMR spectroscopy. ^1H and ^{13}C

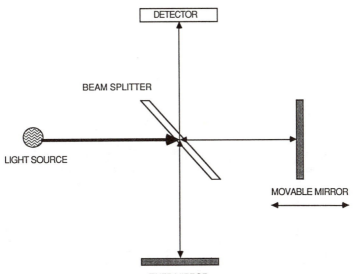

Figure 6.   Schematic diagram of an interferometer as used in FTIR instruments.

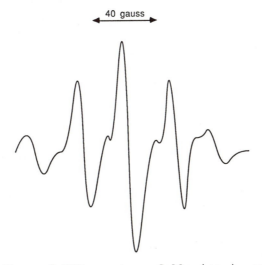

Figure 7.   Form of ESR spectrum following irradiation of poly(methyl methacrylate) at room temperature (after 13). (Reproduced with permission from Ref. 13. Copyright 1985 Cambridge University Press.)

Table I. Information from Spectroscopic Methods

Spectroscopic Technique	Information Available
Electronic (Absorption, fluorescence, phosphorescence)	Chromophore composition, conformation, excited state behaviour, polymer mobility
Vibrational (IR, Raman)	Chemical structure, tacticity, conformation, chemical modification
Magnetic resonance (ESR, NMR)	Chemical structure, tacticity, conformation, polymer mobility (NMR); Radical, triplet state structure and behaviour (ESR)

nuclei are the most useful nuclei in polymers for study by
NMR methods (1).

The value of the magnetic field strength at which
resonances occur for a particular nuclei is influenced by
the chemical environment. The shift in absorption peaks
in NMR spectra arise because the nuclei are shielded from
the external field to differing extents by the diamagnetic
behaviour of surrounding molecular electrons. The
'chemical shift' of various resonance peaks can thus aid
identification of the functional group containing the
nuclei. In addition, hyperfine splittings of the peaks
are observed which may be attributed to interactions
between magnetic nuclei in different parts of the
molecule. Analysis of the hyperfine splitting structure
can thus provide additional information concerning the
local chemical structure. A detailed discussion of the
analysis of NMR spectra is beyond the scope of this
Chapter and the reader is referred elsewhere for further
descriptions (16,18). The spectra obtained from polymers
can be quite complicated, although ^{1}H NMR spectra can
provide important information including details of polymer
tacticity (1).

The introduction of additional techniques such as
Pulsed Fourier Transform NMR spectroscopy (PFT-NMR) has
considerably increased the sensitivity of the method,
allowing many magnetic nuclei which may be in low
abundance, including ^{13}C, to be studied. The additional
data available from these methods allow information on
polymer structure, conformation and relaxation behaviour
to be obtained (1,18,20).

In respect of radiation effects in polymers, the
primary application of NMR spectroscopy is in chemical
analysis to determine the changes in chemical structure
which may occur on exposure.

Conclusion

The variety of spectroscopic methods now available
can be used to provide considerable information on
radiation effects on polymeric materials. These
applications are summarized in Table I. Improvements in
instrumentation and data analysis procedures are
continuing and the development of new spectroscopic
techniques promise new insights into polymer structure and
behaviour.

Acknowledgment
The author acknowledges the assistance of Dr. S. Bigger
in the preparation of the manuscript.

Literature Cited

1.   Klopffer, W. Introduction to Polymer Spectroscopy;
     Springer-Verlag: Berlin, Heidelberg, 1984
2.   Hummel, D.O (ed.) Polymer Spectroscopy; Verlag
     Chemie, Weinheim, 1974
3.   Iving, K.J.(ed.) Structural Studies of Macromolecules
     by Spectroscopic Methods; Wiley, London, 1976
4.   Phillips, E.(ed.) Polymer Photophysics; Chapman and
     Hall, London, 1985
5.   Ghiggino, K.P.; Phillips, D.; Roberts, A.J. Adv.
     Polym. Sci. 1981, 40, 69
6.   Frank, C.W.; Semerak, S.N. Adv. Polym. Sci. 1984, 54,
     31
7.   Ghiggino, K.P.; Bigger, S.W.; Smith, T.A.; Skilton,
     P.F. and Tan, K.L. In Photophysics of Polymers;
     Hoyle, C.E. and Torkelson, J.M. (eds.); ACS Symp Ser.
     No. 358, Am. Chem. Soc., Washington, 1987; Chapter 28
8.   Zbinden, R. Infrared Spectroscopy of High Polymers;
     Academic Press, New York, 1964
9.   Pointer, P.C.; Coleman, M.M.; Koenig, J.L. The Theory
     of Vibrational Spectroscopy and its Application to
     Polymeric Materials; Wiley, New York, 1982
10. Siester, H.W.; Holland-Moritz, K. Infrared and Raman
     Spectroscopy of Polymers; Marcel-Dekker, New York,
     1980
11. Koenig, J.L. Adv. Polym. Sci. 1984, 54, 87
12. Hummel, D.O.; Scholl, F. Atlas der Polymer-und
     Kunstsoffanalyse 2nd Ed.; Verlag Chemie, Munchen,
     1978
13. Grassie, N.; Scott, G. Polymer Degradation and
     Stabilization; Cambridge University Press, Cambridge,
     1985
14. Atkinson, G.H. Time-Resolved Vibrational
     Spectroscopy; Academic Press, New York, 1983
15. Atkinson, G.H. Adv. Infrared and Raman Spectroscopy
     1981, 2, 1
16. McLauchlan, K.A. Magnetic Resonance; University
     Press, Oxford, 1976
17. Ranby, B.; Rabek, J.F. ESR Spectroscopy in Polymer
     Research; Springer, Berlin, 1977
18. Bovey, F.A. High Resolution NMR of Macromolecules;
     Academic Press, New York, 1972
19. Sohma, J.; Sakagudri, M. Adv. Polym. Sci. 1976, 20, 1
20. von Meerwall, E.D. Adv. Polym. Sci. 1984, 54, 1
21. Ghiggino, K.P.; Chemistry in Australia 1987, 54, 450

RECEIVED July 13, 1988

Chapter 4

Flash Photolysis of Aromatic Diisocyanate-Based Polyurethanes

Charles E. Hoyle, Young G. No, and Keziban S. Ezzell

Department of Polymer Science, University of Southern Mississippi, Southern Station Box 10076, Hattiesburg, MS 39406-0076

The laser flash photolysis of aromatic diisocyanate based polyurethanes in solution provides evidence for a dual mechanism for photodegradation. One of the processes, an N-C bond cleavage, is common to both TDI (toluene diisocyanate) and MDI (methylene 4,4'-diphenyldiisocyanate) based polyurethanes. The second process, exclusive to MDI based polyurethanes, involves formation of a substituted diphenylmethyl radical. The diphenylmethyl radical, which readily reacts with oxygen, is generated either by direct excitation (248 nm) or indirectly by reaction with a tert-butoxy radical produced upon excitation of tert-butyl peroxide at 351 nm.

The photochemical processes responsible for the ultimate degradation and destructive failure of polyurethanes based on aromatic diisocyanates have been investigated extensively for the past twenty-five years by a large number of research groups (1-7). Schemes I and II summarize proposed pathways for photodegradation of polyurethanes based on toluene diisocyanate (mixture of 2,4-TDI and 2,6-TDI isomers) and methylene 4,4'-diphenyldiisocyanate (MDI). Scheme I for TDI based polyurethanes (exemplified by the 2,4 isomer) depicts a traditional photo-Fries type rearrangement (4). Scheme II, however, shows a dual mechanism for photodegradation of MDI based polyurethanes recently proposed by Gardette and Lemaire (7), i.e., the traditional photoinduced N-C bond cleavage and subsequent reactions as well as a path for formation of hydrogen peroxide on the central methylene carbon of the bisarylcarbamate moiety with the presumption of further reaction to quinoid type products.

In order to characterize the intermediates leading to the photo-Fries/cleavage and hydroperoxide products shown in Schemes I and II, laser flash photolysis measurements of solutions of both MDI and TDI based polyurethanes were conducted. The results from this study are interpreted by comparison with transient spectra of an aryl monocarbamate and the bispropyl carbamate of MDI. In addition, a dimethylsilicon analog of the MDI bispropyl carbamate is used to

0097-6156/89/0381-0043$06.00/0
© 1989 American Chemical Society

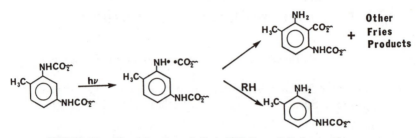

SCHEME I – Photolysis of 2,4-TDI Based Polyurethanes

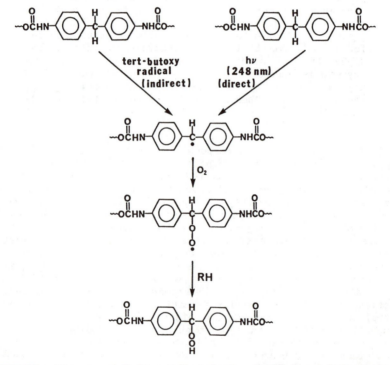

SCHEME II – Direct and Indirect Photolysis of MDI Based Polyurethanes

confirm the mechanism leading to hydroperoxide formation. The transient spectral analysis in this paper supports anilinyl/carboxyl radical formation by N-C bond cleavage. Finally, evidence is presented for diphenylmethyl radical formation in MDI based carbamates.

EXPERIMENTAL

Materials. Methylene 4,4'-diphenyldiisocyanate (MDI, Mobay) was recrystallized from cyclohexane. Toluenediisocyanate (TDI—represents mixture of 2,4- and 2,6-isomers in 80/20 ratio), p-toluidine (Aldrich) and aniline (Aldrich) were purified by vacuum distillation before use. Diphenylmethane, tert-butyl peroxide (TBP), 4-bromoaniline, butyl lithium in hexane, and ethyl chloroformate, were obtained from Aldrich and used as received. Spectrograde tetrahydrofuran (THF) and benzene from Burdick and Jackson were used as received. Poly(tetramethylene ether glycol) with MW 1000 was obtained from polysciences and dehydrated under a rough vacuum at 50 °C for 24 h.

Preparation of model compounds and polyurethanes. A procedure for the preparation of the non-silicon containing monocarbamate and biscarbamate models has been reported previously ($\underline{8}$). In order to obtain the silicon containing model compound, 5.0 g of bis(4-aminophenyl) dimethyl silane [synthesized according to Pratt, et.al ($\underline{9}$)] was added to 7.0 g ethyl chloroformate at room temperature. A salt immediately formed. The mixture was refluxed for 15 min. The solution was then cooled, vacuum filtered, and recrystallized from ethanol to yield 1.0 g of a white powder: MP 162-3 °C; Anal. $C_{20}H_{26}O_4N_2Si$ Calc. C, 62.17; H, 6.70; N, 7.23; Found C, 62.01; H, 6.79; N, 7.42.

Both of the simple polyurethanes (TDI-PU and MDI-PU) were synthesized according to a well known solution polymerization technique ($\underline{10}$). The polyurethane elastomer (MDI-PUE) was prepared by a pre-polymer method ($\underline{11}$).

Instrumentation

The laser flash photolysis unit consists of a Lumonics HyperEx Excimer Laser photolysis source, an Applied Photophysics xenon lamp/monochromator/PMT/auto-offset probe, a Tektronix 7912 transient digitizer, a micro-PDP 11 computer from Digital Equipment Corporation, and an Applied Photophysics control unit. The laser was operated in the charge on demand mode and the xenon lamp source (right-angle arrangement) was momentarily pulsed to achieve a high intensity which was flat for about 200 μs. The laser was operated at either 248 nm (KrF) or 351 nm (XeF). Nominal outputs were 80 mJ/pulse at 248 nm and 60 mJ/pulse at 351 nm. UV absorption spectra of transient intermediates were constructed point-by-point from decay plots taken at specified wavelength intervals. For the kinetic decay studies, the data were analyzed using a software package from Applied Photophysics.

RESULTS AND DISCUSSION

In order to demonstrate the use of laser flash photolysis in elucidation of the MDI based polyurethane photolysis mechanism, three polyurethanes, two aryl biscarbamate models, an aryl monocarbamate model, and an aromatic amine were selected. Two of the polyurethanes are based on MDI while the third is based on TDI (mixture of 2,4 and 2,6 isomers in 80/20 ratio). The MDI based polyurethanes all have the same basic carbamate repeat unit. The MDI elastomer (MDI-PUE) is soluble in tetrahydrofuran (THF). The simple polyurethane (MDI-PU) based on MDI and 1,4-butanediol is used in the tert-butoxy abstraction reactions since it does not contain a polyether backbone. (See page 47 for structures of polymers and models.)

The results and discussion section is divided into two parts. The first part deals with direct laser flash photolysis of the MDI-PUE polymer and appropriate small molecule models. The transient spectra generated by direct excitation of the polyurethane are interpreted by consideration of the primary photochemical reactions of the carbamate moiety. The second part describes results obtained by production of a radical transient species which is capable of abstracting labile hydrogens from the polyurethane. This latter procedure represents an alternative method for production of the transient species which were obtained by direct excitation.

Laser Flash Photolysis at 248 nm of TDI-PU, MDI-PUE, and Model Compounds. Figures 1 and 2 show the transient absorption spectra of MDI-PUE (5.5 X 10^{-3} g/dL) and TDI-PU (2.3 X 10^{-3} g/dL) in THF at a 2.0 μs delay after pulsing with a krypton fluoride excimer laser (λ_{ex}=248 nm) in air and nitrogen saturated samples. Both spectra have common peaks in nitrogen saturated solutions (shown by arrows) at 310 nm, 330-360 nm (broad), and above 400 nm (broad, diffuse absorbance). The MDI-PUE sample has an additional and quite distinctive peak at 370 nm. In the presence of air, the peak at 370 nm for MDI-PUE is completely extinguished, while the sharp peaks at 310 nm for TDI-PU and MDI-PUE and the broad band above 400 nm are only marginally quenched by oxygen.

It should be noted that at the excitation wavelength employed, the absorbance of MDI-PUE is 1.1 while the absorbance of the solvent THF is 0.3. This is a condition dictated by polymer solubility considerations and choice of excitation wavelength. We are confident that the spectral results for the photolysis at 248 nm are derived from radicals generated by direct excitation, as opposed to radical abstraction by solvent radicals, since the kinetic curves indicate no delay in radical formation of the transients.

In order to interpret the results for MDI-PUE and TDI-PU, the laser flash photolysis measurements of several model systems were performed. The transient spectra of the p-toluidinyl radical, recorded upon laser flash photolysis (λ_{ex}=248 nm) of p-toluidine (1.4 X 10^{-4} M in THF), has a distinct maximum at approximately 310 nm and a broad, diffuse absorbance above 400 nm (Figure 3). The results for p-toluidine are in agreement with previously reported spectra for anilinyl type radicals (12,13). Comparing the transient spectra for p-toluidine with TDI-PU and MDI-PUE, it is quite obvious that the p-toluidinyl radical in THF (Figure 3) is essentially identical to the 300-330 nm and > 400 nm portions of the transient spectra of MDI-PUE and TDI-PU in Figures 1 and 2.

POLYMERS

MDI-PU

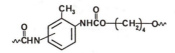

TDI-PU

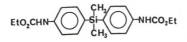

MDI-PUE

MDI + 1,4-Butanediol +
Poly(tetramethylene ether glycol) (MW=1000)

MODELS

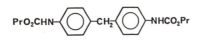

BP-MDI   **p-Toluidine**

SiMe₂-MDI   **Propyl N-tolylcarbamate**

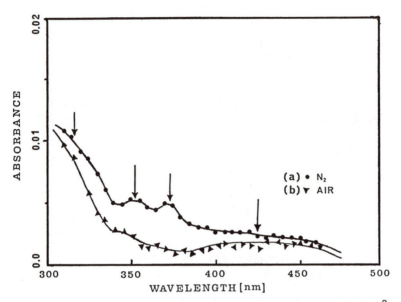

Figure 1. Transient absorption spectra (2.0 μs) of 5.5 X 10^{-3} MDI-PUE in THF (a) Nitrogen saturated (b) Air saturated (λ_{ex}=248 nm) (Reproduced from Ref. 13.   Copyright 1988 ACS).

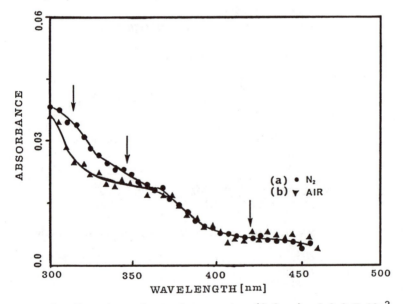

Figure 2. Transient absorption spectra (2.0 μs) of 2.3 X 10^{-3} g/dL TDI-PU in THF (a) Nitrogen saturated (b) Air saturated (λ_{ex}=248 nm).

Before assigning the 310 nm and 400 nm bands in the MDI–PUE and
TDI–PU transient spectra (Figures 1 and 2), the laser flash
photolysis of propyl N–tolylcarbamate (1.3 X 10^{-4} M) was recorded in
THF ($_{ex}$=248 nm). The transient spectrum of propyl N–tolylcarbamate
is very similar to the spectrum reported for propyl N–phenylcarbamate
(13). For propyl N–tolylcarbamate, the peaks at 310 nm, 330–360 nm
and > 400 nm (Figure 4) are identical to three of the four peaks in
the transient spectrum of MDI–PUE and the three peaks in TDI–PU.
Thus, by analogy with p–toluidine, the 310 nm (sharp peak) and 400–
450 nm band in the transient spectrum of propyl N–tolylcarbamate can
be attributed to the p–toluidinyl radical. These results are readily
translated into assignment of the 310 nm and 400–450 nm bands in the
MDI–PUE and TDI–PU transient spectra (Figures 1 and 2) to p–
toluidinyl type radicals. Anilinyl type radicals have been
previously implicated as an intermediate in the photolysis of aryl
carbamates and their presence is certainly expected (4–6).

The broad band absorbance between 330 and 360 nm may be due, at
least in part, to formation of the ortho photo–Fries product.
Contributions from other species may also be important in this region
and final assignment depends on kinetic studies in progress.

The final peak under consideration in the transient spectrum
(Figure 1) of MDI–PUE is the sharp peak at 370 nm which is quite
sensitive to oxygen. Figure 5a shows the transient absorbance
spectrum (λ_{ex}=248 nm) of the bispropyl carbamate of MDI (designated
BP–MDI) in dichloromethane (2.7 X 10^{-5} M). (We recently reported a
similar spectrum for BP–MDI in THF (13)). As in the case of MDI–PUE
(Figure 1), the transient spectrum of BP–MDI in Figure 5a has bands
at 310 nm (sharp peak), 330–360 nm (broad peak), 370 nm (sharp peak)
and above 400 nm (weak, diffuse broad band absorbance). Only the
peak at 370 nm is readily quenched by oxygen. From the
investigations of Porter and Windsor (14), it is well known that
photolysis of diarylmethanes results in a C–H bond cleavage and
formation of diarylmethyl radicals. In the case of diphenylmethane,
the diphenylmethyl radical has a very sharp absorbance maximum at ~
330–360 nm. Substitution at the para positions, such as is the case
with BP–MDI and the biscarbamate group in the MDI–PUE polymer, would
be expected to red shift the absorption maximum. Thus, the peak at
370 nm in Figures 1 and 5a can be tentatively postulated to arise
from the absorbance of a 4,4'–disubstituted diphenylmethyl radical.

In order to provide corroborative evidence for the assignment of
the 370 nm peak to a diphenylmethyl radical, a diphenyl dimethyl
silicon analog of the BP–MDI model compound was prepared (see
structure for this compound which is designated SiMe$_2$–MDI).
Obviously, formation of a diphenylmethyl radical upon photolysis of
SiMe$_2$–MDI is impossible. Accordingly, the transient absorption
spectrum recorded for the laser flash photolysis (λ_{ex}=248 nm) of
SiMe$_2$–MDI (Figure 5b) in dichloromethane (2.6 X 10^{-5} M) has peaks at
310 nm (sharp), 330–360 nm (broad), and above 400 nm (broad, weak
diffuse), but no sharp peak at 370 nm. Apparently, all of the
structural features exhibited by the SiMe$_2$–MDI transient spectrum are
found in the transient spectrum of BP–MDI, save the 370 nm peak.
This certainly provides additional evidence for assigning the 370 nm
peak to a substituted diphenylmethyl radical.

The next section deals with the use of independently generated
tert–butoxy transient species which are capable of abstracting

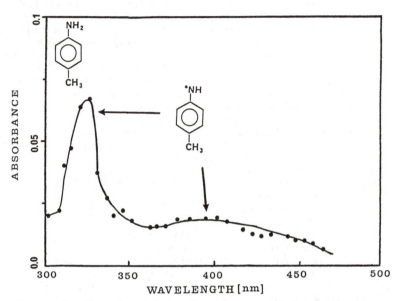

Figure 3. Transient absorption spectrum (2.0 μs) of
1.4 X 10⁻⁴ M p-toluidine in nitrogen saturated THF (λ$_{ex}$=248 nm).

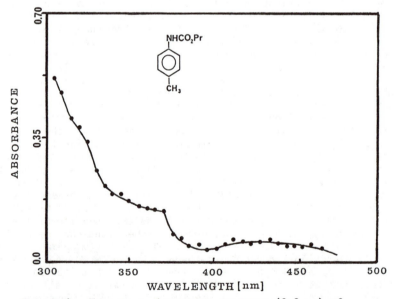

Figure 4. Transient absorption spectrum (2.0 μs) of
1.3 X 10⁻⁴ M propyl N-tolylcarbamate in nitrogen saturated THF
(λ$_{ex}$=248 nm).

hydrogens from the polyurethanes and the model compounds. The results derived from the indirect method of generating transients will be used to substantiate the findings by direct photolysis.

Laser Flash Photolysis at 351 nm of tert-Butyl Peroxide/Benzene Solutions Containing MDI-PUE and Model Compounds. Photolysis of tert-butyl peroxide (TBP) results in a highly efficient production of tert-butoxy radicals. It has recently been shown (15) that tert-butoxy radicals generated by the laser flash photolysis of TBP can rapidly extract hydrogen atoms from appropriate substrates such as aniline and diphenylamine (Scheme III).

The results of an experiment for the laser flash photolysis (λ_{ex}=351 nm) of a 6.0 X 10^{-2} M solution of diphenylmethane in a 60/40 mixture of TBP and benzene (Figure 6) shows a distinct absorbance peak maximum at ~ 340 nm characteristic of the unsubstituted diphenylmethyl radical. The results in Figure 6 illustrate the utility of TBP in indirect generation of diphenylmethyl radicals.

The laser flash photolysis (λ_{ex}=351 nm) of a TBP/BP-MDI solution in benzene (Figure 7) yields a transient spectra with distinct maximum at 370 nm which can most likely be attributed to a substituted diphenylmethyl radical. (Similar results are obtained in other solvents such as DMF). No detectable transient species were generated above 350 nm by the laser flash photolysis (λ_{ex}=351 nm) of the 60/40 mixture of TBP and benzene alone. Results for the TBP/MDI-PU (7.0 X 10^{-2} g/dL) system in Figure 8 show, as in the case of the model BP-MDI (Figure 7), that the transient spectrum of MDI-PU obtained indirectly through tert-butoxy radicals has a maximum at 370 nm. This provides additional support for assignment of the transient species responsible for the 370 nm absorbance to a diphenylmethyl radical.

CONCLUSIONS

The results in this paper support an N–C bond cleavage mechanism (Schemes I and II) for the photolysis of both TDI and MDI based polyurethanes. The substituted anilinyl radicals observed no doubt are formed by diffusion from a solvent cage after the primary N–C bond cleavage. Although not specifically shown in this paper, the reported photo-Fries products (6) are probably formed by attack of the carboxyl radical on the phenyl ring before radical diffusion occurs. The solvent separated anilinyl radicals rapidly abstract hydrogens from the solvent to give the reported aromatic amine product (6).

For MDI based polyurethanes we have provided evidence for formation of a diphenylmethyl radical by direct excitation (248 nm) of the carbamate moiety as well as hydrogen abstraction by a tert-butoxy radical which is produced by excitation (351 nm) of tert-butyl peroxide. The diphenylmethyl radical readily reacts with oxygen. A proposed mechanism which accounts for the production (direct or indirect) and subsequent reaction with oxygen of the diphenylmethyl radical is shown in Scheme IV. The hydrogen peroxide product depicted in Scheme IV has been previously identified by FT-IR (7); we have simply provided a plausible mechanism for its formation.

Finally, the use of TBP to produce photolytically the tert-butoxy radical has not only proven to be an excellent mechanistic

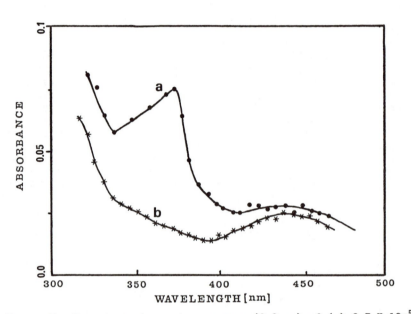

Figure 5. Transient absorption spectra (2.0 μs) of (a) 2.7 X 10⁻⁵ M BP-MDI in nitrogen saturated dichloromethane (λ_{ex}=248 nm) (b) 2.6 X 10⁻⁵ M SiMe₂-MDI in nitrogen saturated dichloromethane (λ_{ex}=248 nm).

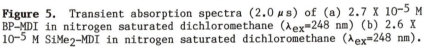

SCHEME III – Photolysis of TBP

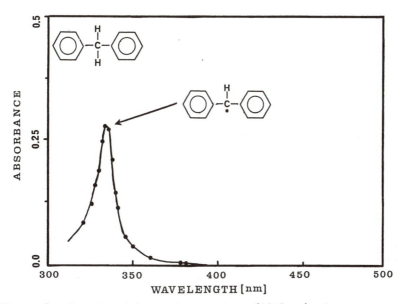

Figure 6. Transient absorption spectrum (12.0 μs) of
6.0 X 10^{-2} M diphenylmethane in a nitrogen saturated TBP/C$_6$H$_6$ (60/40)
solution (λ_{ex}=351 nm).

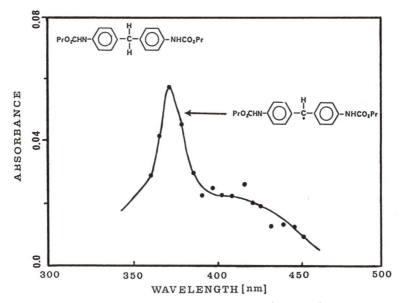

Figure 7. Transient absorption spectrum (10.0 μs) of
3.5 X 10^{-3} M TBP/BP–MDI in a nitrogen saturated TBP/C$_6$H$_6$ (60/40)
solution (λ_{ex}=351 nm).

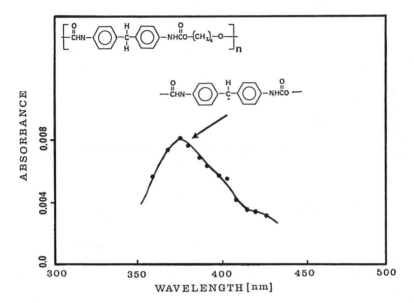

Figure 8. Transient absorption spectrum (15.0 μs) of 7.0 X 10^{-2} g/dL MDI-PU in a nitrogen saturated TBP/DMF (60/40) solution (λ_{ex}=351 nm).

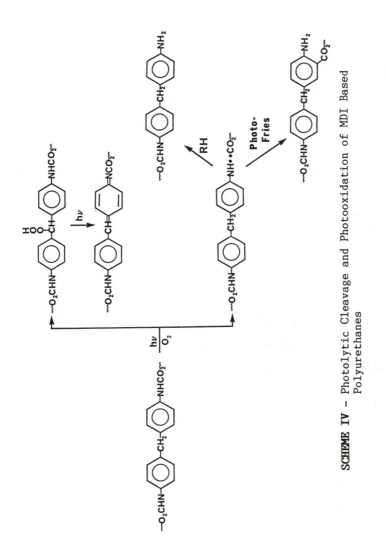

SCHEME IV – Photolytic Cleavage and Photooxidation of MDI Based Polyurethanes

tool for independently generating key radicals, but it also serves as a model for peroxide impurities which may be present in actual polyurethanes. Such peroxide impurities may well be introduced into polyurethanes during processing or upon the initial stages of exposure to terrestrial radiation.

Future work will be concerned with quantitative measurement of hydrogen abstraction rates of labile hydrogens in the carbamate moieties of several aromatic diisocyanate based polyurethanes. It is expected that experimental conditions will alter significantly the hydrogen abstraction rate. Emphasis will also be placed on measurement of transient intermediates in polyurethane films. Finally, extensive laser flash photolysis experiments will be conducted on polyurethanes based on both 2,4-toluenediisocyanate and 2,6-toluenediisocyanate. Preliminary data suggest that the placement of the methyl substituent can alter the nature of the transient intermediates formed.

Acknowledgments

This work was sponsored in part by the Office of Naval Research. In addition, acknowledgment is made to the donors of the Petroleum Research Fund, administered by the American Chemical Society, for partial support of this research. Acknowledgment is also made to NSF for assistance in purchasing the laser flash photolysis unit (Grant CHE-8411829-Chemical Instrumentation Program).

REFERENCES

1. Schollenberger, C. S.; Stewart, F. D. J. Elastoplastics, 1972, 4, 294.
2. Schollenberger, C. S.; Stewart, F. D. Advances in Urethane Science and Tchnology, Vol. II, ed. by K. C. Frisch and S. L. Reegen, Technomic Publishing, Connecticut, USA, 1973.
3. Nevskii, L. V.; Tarakanov, O. G.; Beljakov, V. K. Soviet Plastics, 1966, 7, 45.
4. Beachell, H. C.; Chang, I. L. J. Polym. Sci., Part A, 1972, 10, 503.
5. Osawa, Z; Cheu, E. L.; Ogiwara, Y. J. Polym. Sci., Letter Ed., 1975, 13, 535.
6. Hoyle, C. E.; Kim, K. J. J. Polym. Sci., Polym. Chem. Ed., 1986, 24, 1879.
7. Gardette, J. L.; Lemaire, J. J. Polym. Deg. and Stab., 1984, 6, 135.
8. Hoyle, C. E.; Herweh, J. E. J. Org. Chem., 1980, 11, 2195.
9. Pratt, J. R.; Massey, W. D.; Pinkerton, F. H.; Thames, S. F. J. Org. Chem., 1975, 40, 1090.
10. Lyman, D. J. J. Polym. Sci., 1960, 45, 409.
11. Saunders, J. H.; Frisch, K. C. "Polyurethanes: Chemistry and Technology," Vol. XVI, 1962, Interscience Publishers, New York, p. 273.
12. Land, E. J.; Porter, G. Trans. Faraday Soc., 1963, 59, 2027.
13. Hoyle, C. E.; No, Y. G.; Malone, K. G.; Thames, S. F. Macromolecules, accepted.
14. Porter, G.; Windsor, W. R. Nature, 1957, 180, 187.
15. Platz, M. S.; Leyva, E.; Niu, B.; Wirz, J. J. Phys. Chem.,1987 91, 2293.

RECEIVED September 2, 1988

Chapter 5

Photophysics of Hydroxyphenylbenzotriazole Polymer Photostabilizers

Kenneth P. Ghiggino, A. D. Scully, and S. W. Bigger

Department of Physical Chemistry, University of Melbourne, Parkville 3052, Australia

The efficiency of the energy relaxation processes occurring in hydroxyphenylbenzotriazole polymer photostabilizers following absorption of ultraviolet (UV) radiation has been investigated and found to depend strongly on the stabilizer structure and the nature of the solvent or polymer substrate. Steady-state absorption and fluorescence, and picosecond fluorescence spectroscopy measurements were carried out on a number of these compounds. The very rapid excited-state intramolecular proton transfer (ESIPT) process responsible for the exceptional photostability of these derivatives is influenced by the hydrogen-bonding properties and polarity of the surrounding medium. The ESIPT process is prevented effectively in those molecules in which the intramolecular hydrogen bonding is disrupted by hydrogen-bonding or steric interactions with the solvent or polymer environment.

Most synthetic and natural polymers degrade when exposed to solar ultraviolet (UV) radiation (1-5). In synthetic polymers degradation is generally caused by the presence of photosensitive impurities and/or abnormal structural moieties which are introduced during polymerization or in the fashioning of the finished products. The presence of groups such as ketones, aldehydes, peroxides and hydroperoxides are implicated in polymer degradation (1-5).

A major cause of polymer degradation is free-radical chain oxidation which results in the cleavage of the polymer backbone, unsaturation, crosslinking or the formation of small molecular fragments (1-5). Other photodegradative processes include Norrish-type cleavage of carbonyl groups and other unsaturated structural groups

0097–6156/89/0381–0057$06.75/0
© 1989 American Chemical Society

as well as oxidative photodegradation due to the presence
of singlet oxygen (1-5). These irreversible chemical
processes lead to aesthetic changes such as discoloration
and cracking as well as to the breakdown in mechanical and
electrical properties of the polymer.

Most commercial polymers are protected against UV
degradation by incorporating additives into the bulk
polymer. For example, excited-state quenchers stabilize
polymeric materials by the quenching of species which are
excited by the absorption of light, whereas free-radical
scavengers inhibit chain propagation processes by the
removal of free radicals or by decomposing hydroperoxides to produce non-reactive species (1-5). Two
important groups of additives are UV screens and UV
absorbers. Ultraviolet screens, such as the oxides of
titanium, iron and chromium, protect the polymer by
reflecting harmful radiation thus preventing its
absorption by chromophores in the polymer (1-5).

Ultraviolet absorbers preferentially absorb damaging
UV radiation and this energy is dissipated harmlessly by
various non-radiative and radiative processes which occur
in the stabilizer. For a UV absorber to be an effective
stabilizer it must have a large extinction in the
wavelength range over which the polymer is most
susceptible to photodegradation. The wavelengths of
maximum photosensitivity for some commercially important
polymers are: 318 nm for poly(styrene) (PS), 310 nm for
poly(vinylchloride) (PVC), 295 nm for the polycarbonates
and 290 to 315 nm for poly(methylmethacrylate) (PMMA) (4).
Ultraviolet stabilizers must be compatible with the
polymer substrate, be capable of surviving the processing
conditions and should not be exuded or leached readily
from the polymer.

It is now possible to study the very rapid initial
excited-state processes that occur in polymers and polymer
additives following irradiation with UV light using both
steady-state and laser-based picosecond fluorescence
spectroscopy. One application of these techniques is in
the study of the photophysical processes occurring in
polymers containing photostabilizers such as the
salicylates (6,7), o-hydroxyphenyl-s-triazines (8),
o-hydroxybenzophenones (7), and o-hydroxyphenylbenzo-
triazoles ("benzotriazoles") (9-19). These compounds have
found wide application as UV absorbers and photostabilizers for polymeric materials although it is known
that their effectiveness varies markedly depending on the
nature of the substrate. In particular, recent studies
indicate that the photophysics of these and related
compounds is influenced strongly by the molecular
environment of the photostabilizer (11,12,14,16-20).

The 2-(2'-hydroxyphenyl)benzotriazole derivatives,
for example, are capable of forming intramolecular
hydrogen bonds between a nitrogen atom on the triazole
ring and the phenolic proton. This hydrogen bond
facilitates the occurrence of a very rapid excited-state

intramolecular proton transfer (ESIPT) process following
the initial excitation of the stabilizer from its ground
state (S_0) to the first excited singlet state (S_1). The
resulting proton-transferred (zwitterionic) form of the
molecule (S_1') is believed to dissipate the excitation
energy rapidly by internal conversion (IC), a radiation-
less and non-degradative process. This deactivation
pathway gives rise to the high degree of photostability of
these molecules (9-19). The short excited-state lifetimes
of these molecules also decreases their likelihood of
initiating degradative processes in the substrate and this
feature, combined with their high extinction coefficients,
makes them suitable as UV absorbers. Figure 1 is a
schematic diagram that summarizes the photophysical
processes that occur in these types of stabilizers
following the absorption of UV light.
        Recently, the photophysics of the widely used UV
absorber 2-(2'-hydroxy-5'-methylphenyl)benzotriazole (TIN)
(10-12,14,16,18,19) and its sulfonated derivative, sodium
2-(2'-hydroxy-5'-methylphenyl)benzotriazole-3'-sulfonate
(TINS) (14,17) (see Figure 2), were studied in solution
and it was proposed that the ground-state conformation
adopted by each of these molecules is sensitive to the
molecular environment.
        In non-polar solvents TIN is believed to exist
mainly in an intramolecularly hydrogen-bonded form which
is non-fluorescent at room temperature (9,12,16).
Fluorescence emission ($\lambda_{max} \approx 630$ nm) from the proton-
transferred form of the molecule is observed in low
temperature, non-polar glasses (9,10,12,16) and in the
crystalline form (15). In polar, hydrogen-bonding
solvents a shorter wavelength fluorescence ($\lambda_{max} \approx 400$ nm)
is observed which is attributed to those molecules which
are intermolecularly hydrogen bonded to the solvent and
which do not undergo ESIPT (10-12,14,16-19).
        The photophysics of some benzotriazole derivatives
in different molecular environments created by various
solvents and polymer substrates is reported in this work.
In particular, various aspects concerning the photophysics
of the o-hydroxyphenylbenzotriazole stabilizers TIN, TINS,
sodium 2-(2'-hydroxy-5'-methylphenyl)benzotriazole-5-
sulfonate (STIN) and sodium 2-(2'-hydroxy-3'-t-butyl-5'-
methylphenyl)benzotriazole-3'-sulfonate (t-Bu-STIN) are
discussed.

Experimental

Materials. The structures of various o-hydroxyphenyl-
benzotriazoles which have been studied extensively by our
reseach group are shown in Figure 2. Each of these
compounds was synthesized, purified and supplied by Dr. B.
Milligan and Mr. P. J. Waters of the CSIRO Division of
Protein Chemistry. The derivative 2-(2'-methyl-5'-
methoxyphenyl)benzotriazole (MeTIN) was supplied by Dr. R.
G. Amiet of the Royal Melbourne Institute of Technology.

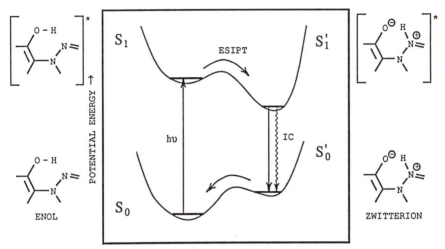

HYDROGEN/OXYGEN DISTANCE →

Figure 1.   Schematic diagram of the ESIPT process.

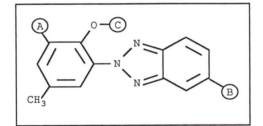

BENZOTRIAZOLES	A	B	C
TIN	H	H	H
TINS	SO_3Na	H	H
STIN	H	SO_3Na	H
t-Bu-STIN	t-Bu	SO_3Na	H
MeTIN	H	H	CH_3

Figure 2.   Structures of the benzotriazoles studied.

Poly(styrene) and PMMA were synthesized from their respective monomers using azobisisobutyronitrile-initiated radical polymerization in benzene. Four freeze-pump-thaw cycles were used to degas the monomer solutions and polymerization was carried out for 48 hours at 60°C. The polymers were purified by multiple reprecipitations from dichloromethane into methanol. Films of these polymers were prepared and found to be free of any fluorescent impurity.

Each solvent used was observed to contain no impurities which fluoresce in the spectral region of interest. All solution concentrations used were in the range 10^{-5} to 10^{-4} M. Polymer films were cast onto quartz plates from either chloroform or dichloromethane solutions containing 4% (wt/wt) of polymer. The films were air dried at room temperature and had an average thickness of 65 ± 10 μm. The absorption spectra of the polymer films were measured using an appropriate PMMA or PS film as the reference.

Methods. Absorption spectra were recorded using an Hitachi model 150-20 spectrophotometer/data processor system. Uncorrected steady-state fluorescence emission spectra were recorded using a Perkin-Elmer MPF-44A spectrofluorimeter. These spectra were collected and stored using a dedicated microcomputer and then transferred to a VAX 11/780 computer for analysis. Fluorescence spectra were corrected subsequently for the response characteristics of the detector (21). Values of the fluorescence quantum yield, ϕ_f, were determined relative to either quinine bisulfate in 1N H_2SO_4 (ϕ_f = 0.546) (22) or a pyrazoline optical brightener contained within a polymer film (ϕ_f = 0.92) (23).

A picosecond laser/streak camera system (see Figure 3) was used to obtain the fluorescence decay profiles. This system uses the third (353 nm) or fourth (265 nm) harmonic of a single pulse from a mode-locked Nd^{3+}/phosphate glass laser (pulse width of approximately 5 ps) to excite the sample. The fluorescence emitted from the samples is detected at right angles to the excitation direction, recorded using a Photochron II streak camera and stored using an optical multichannel analyser (OMA). Appropriate cut-off and band-pass filters were used to isolate the emission to be studied. Further details relating to the picosecond laser/streak camera system are given elsewhere (24). Data from the OMA are then transferred to a VAX 11/780 computer for analysis. Values of the fluorescence decay time, τ_f, were calculated using an iterative, non-linear, least-squares fitting method which is described elsewhere (21) in which reconvolution is necessary due to a finite instrument response function (approx. 30 ps FWHM).

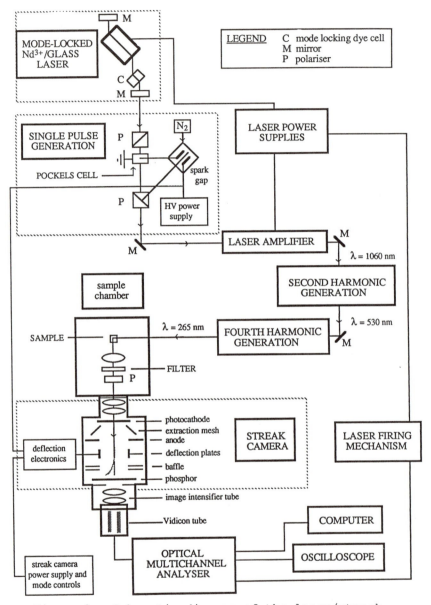

Figure 3.   Schematic diagram of the laser/streak-camera/OMA apparatus.

Results and Discussion

Absorption Spectra. The absorption spectrum of TIN has been studied extensively in solution (4,9,14,16-19) and is found to consist of two absorption bands with maxima at approximately 300 nm and 350 nm (Figure 4). The longer wavelength absorption band is associated with a planar form of the molecule (TIN(*planar*)) and the shorter wavelength band is due primarily to a non-planar form (TIN(*non-planar*)) in which the intramolecular hydrogen bond is disrupted and conjugation between the two ring systems is reduced substantially (4,14-16,17). These assignments are consistent with the general observation of a red-shifting of the absorption spectrum with increasing conjugation due to the decrease in the π - π* energy gap (25) and are supported further by the absorption spectrum of MeTIN in solution. This molecule cannot exist in a planar (intramolecular hydrogen-bonded) form and consequently only the higher energy absorption band is observed.

In the case of benzotriazole compounds which display both absorption bands, the observed spectrum consists of the superposition of the individual spectra that correspond to the two distinct ground-state species. The absorption spectrum of TIN in methanol/dimethylsulfoxide (DMSO) solvent mixtures varies with the composition of the solvent (Figure 5) in a manner which suggests that the proportion of planar and non-planar forms of TIN is solvent dependent.

The absorption spectra of the TIN(*planar*) and TIN(*non-planar*) forms can be resolved using the mathematical least-squares method of Principal Component Analysis (PCOMP) (26-28). In this analysis, the spectra of the two components can be calculated from a series of spectra containing different proportions of each component and no assumptions are made as to the shapes of the curves. A unique solution is obtained only if the two components have zero intensity at different wavelengths, otherwise a range of solutions is obtained. The unique component spectra obtained account for 99.99 percent of the variance of the original data.

The absorption spectrum of each resolved component of TIN in the methanol/DMSO mixtures can be calculated by virtue of the isosbestic point at 290 nm because at this wavelength the extinction coefficients of the two forms are equal (Figure 6). At 300 nm the extinction coefficient of the planar and non-planar forms is 1.5 x 10^4 M^{-1} cm^{-1} and 1.0 x 10^4 M^{-1} cm^{-1} respectively. These values are very similar to those calculated for TINS in acetonitrile/water mixtures (1.8 x 10^4 M^{-1} cm^{-1} and 1.0 x 10^4 M^{-1} cm^{-1} for the planar and non-planar forms respectively) (17).

The PCOMP analysis also enables the relative proportion of molecules in the planar and non-planar forms to be calculated for the different solvent compositions

NORMALIZED ABSORBANCE
(arbitrary units)

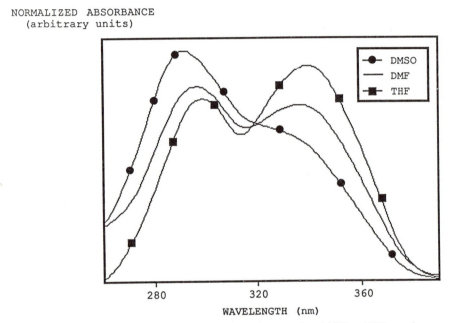

WAVELENGTH (nm)

Figure 4.   Absorption spectra of TIN in DMSO, THF and
N,N-dimethylformamide (DMF).

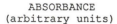

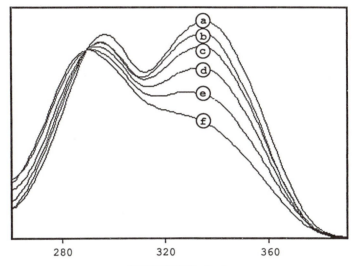

WAVELENGTH (nm)

Figure 5. Absorption spectra of TIN in methanol/DMSO mixtures. Molar percentages of methanol are: (a) 100%, (b) 88%, (c) 73%, (d) 54%, (e) 31% and (f) 0%.

(see Figure 7). The absorption spectrum for each solvent composition is fitted with a linear combination of the two resolved component absorption spectra using a multiple regression analysis (29). The relative proportion of TIN(planar) and TIN(non-planar) in a range of solvents can be estimated using this fitting procedure. Figure 8 shows the fit to the absorption spectrum of TIN in acetonitrile using this technique.

The gradual decrease in the proportion of TIN(non-planar) upon the addition of methanol (Figure 7) seems best explained in terms of a bulk solvent effect. Dimethylsulfoxide, an aprotic solvent, can form strong hydrogen bonds with the phenolic proton of TIN and thus may disrupt the intramolecular hydrogen bond and promote a shift in the ground-state equilibrium towards the non-planar form. Methanol, a protic solvent, may form a solvated complex which appears to favour the formation of the planar structure (17). If the ground-state energies of these solvation complexes are assumed to be similar then the relative populations of the two forms will vary widely and depend on the composition of the solvent mixture.

Table I lists the proportions of the non-planar conformers of TIN and TINS that are present in various solvents and the contribution to the total absorbance made by the non-planar conformer, $\%A_{np}(\lambda)$, at the excitation wavelength, λ, calculated using Equation 1, together with relevant solvent parameters.

$$\%A_{np}(\lambda) = 100 \times \varepsilon_{np}(\lambda) f_{np} / [\varepsilon_{np}(\lambda) f_{np} + \varepsilon_p(\lambda) f_p] \qquad (1)$$

where $\varepsilon_{np}(\lambda)$ and $\varepsilon_p(\lambda)$ are the extinction coefficients of the non-planar and planar conformers respectively and f_{np} and f_p are the fractions of non-planar and planar conformers calculated from the multiple regression analysis.

The results in Table I show that the percentage of the non-planar form varies with both the polarity and hydrogen-bonding strength of the solvent. In particular, this conformation of the molecules is expected to be more polar than the planar form (25) and thus should be stabilized by solvents of greater polarity. However, the hydrogen-bonding strength of the solvent also plays a significant role in determining the relative proportion of each form. In the case of acetonitrile, which is a relatively polar but weakly hydrogen-bonding solvent, approximately 20% of the TIN molecules exist in a non-planar conformation. In contrast, TIN dissolved in DMSO, which is also a highly polar solvent but has a strong hydrogen-bonding propensity, favors the non-planar form. In this solvent approximately 70% of the molecules adopt a non-planar conformation. A substantially higher proportion of TINS molecules in solution adopt a non-planar conformation as compared with TIN. This may be due to competitive intramolecular hydrogen bonding by the

EXTINCTION COEFFICIENT

$(M^{-1} cm^{-1})$

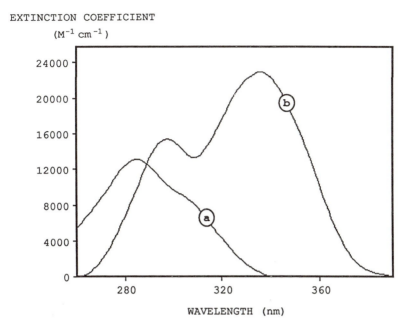

WAVELENGTH (nm)

Figure 6. PCOMP-resolved absorption spectra of: (a) TIN(*non-planar*) and (b) TIN(*planar*).

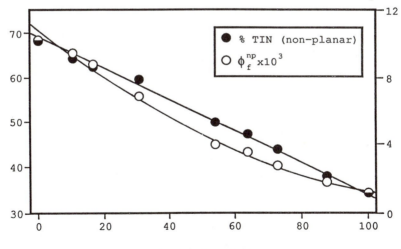

MOLE % METHANOL

Figure 7. Plot of mole percent of TIN(*non-planar*) and ϕ_f^{np} versus mole percent of methanol for methanol/DMSO solvent mixtures.

ABSORBANCE
(arbitrary units)

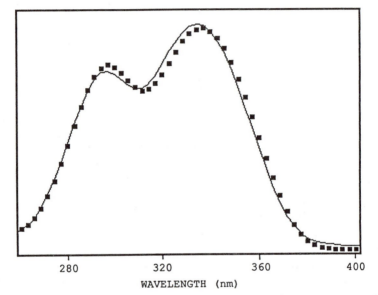

WAVELENGTH (nm)

Figure 8. Fit of PCOMP-resolved absorption spectra of
TIN (solid line) to the absorption spectrum of TIN in
acetonitrile (squares) (correlation coefficient: 0.997).

Table I. Percentage mole fractions of non-planar conformers and their contributions to absorbance at 300 nm for TIN and TINS

SOLVENT	TIN		TINS			
	% non-planar	% A_{np} (300 nm)	% non-planar	% A_{np} (300 nm)	$E_T (30)^{[1]}$ (kJ mol^{-1})	$DN^{[2]}$ (kJ mol^{-1})
carbon tetrachloride	4	3	–	–	136.0	0
diethyl ether	12	8	–	–	144.8	80.4
tetrahydrofuran	13	9	80	69	156.6	83.7
ethyl acetate	13	9	79	68	159.5	71.6
acetonitrile	23	17	97	95	192.6	59.0
ethanol	28	21	75	63	217.3	–
methanol	34	26	70	57	232.3	79.5
water	–	–	57	42	264.0	75.3
N-methylformamide	43	34	–	–	226.5	–
N,N-dimethylformamide	46	36	–	–	183.3	111.3
dimethylsulfoxide	68	59	–	–	188.4	124.7

Notes : [1] Empirical solvent polarity parameter (see Reference 32)
        [2] Solvent hydrogen-bonding strength using Donor Number parameter (see Reference 32)

sulfonate group which is situated in a position *ortho* to
the hydroxy group in the molecule.
    The ground states of the TIN and TINS stabilizers
respond to the influence of the molecular environment in
polymer films in almost the same manner as they do in
solution. The absorption spectra of TIN in PMMA film
(Figure 9) and TIN in PS film are similar to those
observed for TIN in low polarity, non hydrogen-bonding
solvents. A linear combination of the TIN(*planar*) and
TIN(*non-planar*) component spectra from the PCOMP analysis
was used to fit the absorption spectrum of TIN in PMMA.
It is estimated that in this polymer film approximately
20% of the TIN molecules exist in the non-planar form.
The spectrum of TIN in PS is red-shifted compared to its
spectrum in PMMA and so the relative proportion of each
form could not be calculated using this method. However,
the similarity between these two spectra suggests that a
comparable proportion of TIN molecules in PS assume a
non-planar conformation. The absorption spectrum of MeTIN
in a PMMA film consists of a single absorption band (see
Figure 9). This band is similar to that observed for
MeTIN in solution suggesting that MeTIN exists almost
entirely in a non-planar conformation in this polymer.
The proportion of the non-planar form of TINS in films of
poly(vinylalcohol) (PVA) a protic, hydrogen-bonding
polymer and poly(vinylpyrrolidone) (PVP) an aprotic,
hydrogen-bonding polymer, is approximately 10% and 90%
respectively (17). These results demonstrate that the
polymer environment can have a profound effect on the
preferred conformation of the stabilizer and thus
influence its effectiveness in photostabilizing the
substrate.

Fluorescence Spectra. Studies of the crystalline
structure of TIN show that it is a planar molecule with a
strong intramolecular hydrogen bond (15). The large
red-shift observed in the fluorescence emission spectrum
($\lambda_{max} \approx 630$ nm) indicates that a significant change in the
geometry of the excited-state molecule, compared with the
ground state, occurs and this suggests that the emission
arises from the excited state of the proton-transferred
tautomer.
    In methanol/DMSO solvent mixtures the fluorescence
spectrum of TIN ($\lambda_{max} \approx 400$ nm) displays a normal Stokes
shift indicating that this emission arises from a non
proton-transferred, excited state of TIN. The
fluorescence excitation spectrum for this emission
coincides with the absorption spectrum of the resolved
non-planar species suggesting that this conformer is the
ground-state precursor responsible for the observed
emission. As the amount of DMSO in the mixture increases
the fluorescence maximum undergoes a bathochromic shift
from 415 nm in pure methanol to 440 nm in pure DMSO.
    Figure 7 shows the variation of the fluorescence
quantum yield of TIN in different methanol/DMSO mixtures.

The quantum yield values (ϕ_f^{np}) are corrected for the absorption of the excitation light by the non-fluorescent, planar component using Equation 2.

$$\phi_f^{np} = \phi_f^{tot} \; (1 + A_p(\lambda)/A_{np}(\lambda)) \tag{2}$$

where ϕ_f^{tot} is the fluorescence quantum yield calculated using the total absorbance of the solution and $A_p(\lambda)/A_{np}(\lambda)$ is the ratio of the absorbances of the planar and non-planar forms respectively measured at the excitation wavelength. The gradual decrease in the value of ϕ_f^{np} as the mole percent of methanol increases suggests that the quenching of the fluorescence by methanol is a bulk solvent effect.

The fluoresence lifetimes calculated for TIN in low viscosity alcohols are approximately proportional to the solvent viscosity (19) which suggests that in these solvents there is a non-radiative process related to the rotational diffusion mobility of the TIN molecule. The observed extent of the quenching, however, is significantly greater than that expected due to viscosity effects alone and cannot be explained by a collisionally-induced, Stern-Volmer type process involving methanol molecules (25) as the appropriate plot is non-linear.

The values of ϕ_f^{tot} for various benzotriazole compounds in a range of solvents are listed in Table II. Values of the fluorescence quantum yield for TIN and TINS, corrected for the absorbance by their non-fluorescent, planar conformers at the excitation wavelength, are listed in Table III. In all the benzotriazole solutions examined, maximum fluorescence emission was observed at about 400 nm indicating that this emission originates from the non proton-transferred species. This was confirmed by examination of the fluorescence excitation spectrum which corresponds to the absorption spectrum of the non-planar form of the molecule.

The magnitude of the fluorescence quantum yield is determined by the nature of the solvent. The fluorescence quantum yield values are significantly larger in aprotic hydrogen-bonding solvents than in polar, protic hydrogen bonding solvents. In aprotic solvents it is expected that the intramolecular hydrogen bond is replaced to a large extent by intermolecular hydrogen bonds between the solvent and the stabilizer thus disrupting the ESIPT process. This removes an efficient non-radiative pathway and increases the fluorescence yield.

No fluorescence is observed at room temperature from TIN in non-polar solvents such as cyclohexane. In these solvents only the intramolecularly hydrogen-bonded form, which can undergo rapid ESIPT upon excitation, is present. The t-Bu-STIN derivative (see Table II) is very weakly fluorescent in all of the solvents examined. This is attributable to the protection of the intramolecular hydrogen bond from the solvent by the tertiary butyl group which is adjacent to the labile proton.

NORMALIZED ABSORBANCE
(arbitrary units)

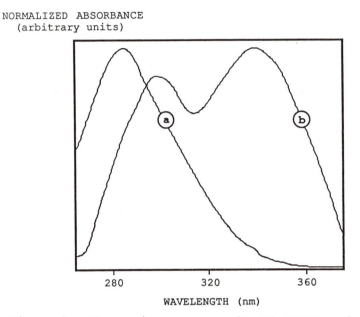

WAVELENGTH (nm)

Figure 9. Absorption spectra of: (a) MeTIN and (b) TIN
in PMMA films.

Table II. Values of fluorescence quantum yield for some
benzotriazoles in a range of solvents

		ϕ_f^{tot} x 10^3		
SOLVENT	TIN	TINS	STIN	t-Bu-STIN
water	–	1.1	0.1	0.04
methanol	0.3	8.7	0.3	0.1
ethanol	0.5	29	0.4	0.1
N-methylformamide	1.0	–	–	–
diethyl ether	1.3	–	–	–
acetonitrile	2.2	160	2.6	0.3
N,N-dimethylformamide	5.6	–	6.0	–
dimethylsulfoxide	6.0	–	5.0	–
ethyl acetate	6.8	168	1.8	0.1
tetrahydrofuran	10	130	2.4	0.3

Table III. Photophysical properties of TIN and TINS in various solvents

SOLVENT	$\phi_f^{np} \times 10^3$	τ_f (ps)	$k_f \times 10^{-6}$ (s^{-1})	$k_{nr} \times 10^{-7}$ (s^{-1})
[1] TIN				
methanol	1.2	20 ± 1	60 ± 15	5000 ± 250
ethanol	2.4	53 ± 1	45 ± 10	1887 ± 36
dimethylsulfoxide	10	400*	25 ± 5	250 ± 20
acetonitrile	13	544 ± 2	24 ± 5	184 ± 7
N,N-dimethylformamide	16	516 ± 25	30 ± 8	194 ± 9
ethyl acetate	76	1212 ± 77	62 ± 16	83 ± 5
tetrahydrofuran	111	1285 ± 49	86 ± 20	78 ± 3
[2] TINS				
methanol	15	64 ± 7	24 ± 7	1563 ± 171
ethanol	46	128 ± 6	36 ± 9	781 ± 37
acetonitrile	168	694 ± 27	24 ± 6	12 ± 1
tetrahydrofuran	188	751 ± 39	25 ± 6	10 ± 1
ethyl acetate	247	797 ± 24	31 ± 7	11 ± 1

* value obtained from Reference 16

The low values of the fluorescence quantum yield obtained for the benzotriazoles in protic, hydrogen-bonding solvents can be explained by the formation of an excited-state encounter complex between the stabilizer and solvent molecules. This allows proton transfer to occur indirectly through the solvent molecules (14,17,19) (see Figure 10). Higher fluorescence yields and longer lifetimes for TIN and TINS in deuterated methanol (14) provide further support for the involvement of the solvent in the non-radiative process. A similar mechanism to this has been proposed to be operative in systems involving 3-hydroxyflavone (20) and 7-azaindole complexes (30). If the relative energies of the ground-state complexes of TIN with methanol and with DMSO are similar then the proportion of molecules in each type of complex will be determined by the relative concentration of each solvent in the immediate vicinity of the TIN molecule. Consequently, behaviour similar to that shown in Figure 7 might be expected.

The fluorescence emission spectrum of TIN in PS and PMMA films is dominated by a blue emission with a maximum at approximately 400 nm. The fluorescence excitation spectrum, monitoring this emission, corresponds to the absorption band of TIN(*non-planar*) obtained from the PCOMP analysis indicating that it is this form that leads to the observed emission.

The value of the quantum yield of fluorescence of TIN in the PMMA film (calculated using the total film absorbance at the excitation wavelength) decreases from 1.2×10^{-3} to 5.0×10^{-4} when the concentration of TIN is increased from 0.07 mole% to 5.0 mole%. This suggests that the TIN molecules are involved in a concentration-dependent, self-quenching process.

The fluorescence emission spectra of TINS in PVA and PVP also show only a single band near 400 nm which is attributable to emission from a non proton-transferred excited state. The similarity between the values of the fluorescence quantum yield, ϕ_f^{np}, for the non proton-transferred form of TINS in PVA and PVP (17) indicates that the PVA polymer is unable to behave in an analogous manner to protic, hydrogen-bonding solvents and suggests that no complexation which can facilitate ESIPT occurs in the excited state as a result of the restricted motion of the PVA chains.

It should be noted that no emission from the zwitterionic form of the proton-transferred tautomer was observed from any of the benzotriazoles studied in the present work. This implies that non-radiative relaxation processes from the excited state of this species are very efficient in all of the solvent and polymer environments studied. Thus no information is available on the effect of the medium polarity on the room-temperature photophysics of the zwitterionic form using fluorescence techniques.

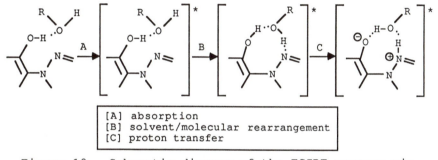

[A] absorption
[B] solvent/molecular rearrangement
[C] proton transfer

Figure 10.   Schematic diagram of the ESIPT process via solute/solvent complexation.

Fluorescence Lifetimes. The fluorescence decay times of TIN in a number of solvents (11,14,16,18,19), low-temperature glasses (12) and in the crystalline form (15) have been measured previously. Values of the fluorescence lifetime, τ_f, of the initially excited form of TIN and TINS in the various solvents investigated in this work are listed in Table III. Values of the radiative and non-radiative rate constants, k_f and k_{nr} respectively, are also given in this table. A single exponential decay was observed for the room-temperature fluorescence emission of each of the derivatives examined. This indicates that only one excited-state species is responsible for the fluorescence in these systems.

The much shorter lifetimes that are observed in protic, hydrogen-bonding solvents compared with aprotic, hydrogen-bonding solvents are consistent with the presence of an additional non-radiative process which is suggested to be ESIPT via a solvent/solute complex (see Figure 10). The longer lifetimes determined for the benzotriazoles in the presence of the larger protic solvent species suggest that orientation of the solvent molecule is an important step in the deactivation mechanism which involves solvent/solute complexation. This is confirmed by a recent study which shows a correlation between solvent viscosity and the fluorescence lifetime of TIN dissolved in various 1-alkanols (19).

The values of the radiative rate constants calculated for TIN and TINS remain reasonably constant for each of the solvents used. This observation is consistent with the assumption that the same excited-state species, namely a non proton-transferred tautomer, is responsible for fluorescence in each case. The significantly larger non-radiative decay rate constants found in alcoholic solvents can be attributed to both molecular reorientation of the solvent molecules and, possibly, rotation of the hydroxy group out of the molecular plane in order to accommodate the solvent bridge required for ESIPT. The lower values of k_{nr} found for TIN and TINS in aprotic hydrogen-bonding solvents suggest that ESIPT does not occur in those molecules that are intermolecularly hydrogen bonded to the solvent. A longer fluorescence lifetime was observed for TIN in tetrahydrofuran (THF) and ethyl acetate than in more polar aprotic solvents suggesting a polarity dependence of the non-radiative pathway of the non-proton transferred, excited-state tautomer. The higher value of k_{nr} for TIN in high polarity, aprotic solvents has been attributed to intersystem crossing in the first excited state (14).

The fluorescence decay time, calculated for the 400 nm emission of TIN in PMMA films, decreases from 1.3 ± 0.2 ns to 0.20 ± 0.02 ns as the concentration of TIN is increased from 0.07 mole% to 1.1 mole% respectively. This is further evidence that the TIN molecules are involved in a concentration-dependent, self-quenching process suggesting that aggregation of the stabilizer molecules occurs at high concentrations in polymer films (31).

The excited-state lifetime calculated for TINS in PVA film is found to be 1.3 ± 0.1 ns compared with 44 ± 4 ps found in the case of water (17). This supports the earlier proposal that complexation, which is proposed to occur in protic, hydrogen-bonding solvents, does not occur in this polymer. In the PVP film an intense fluorescence and a long excited-state lifetime, similar to that found for TINS in PVA, is observed and is consistent with the ESIPT process being prevented in this aprotic medium.

Conclusions

The absorption spectra of the hydroxyphenylbenzo-triazole derivatives in various solvents and polymer films indicate that two ground-state forms of the molecules exist. These species are proposed to be a planar and non-planar form of the stabilizers. The position of the equilibrium between these two forms is affected by both the polarity and the hydrogen-bonding strength of the medium. The blue fluorescence (λ_{max} ≈ 400 nm) observed for these stabilizers originates from an excited-state species in which intramolecular proton transfer is disrupted.

The dependence of the fluorescence quantum yields and lifetimes of these stabilizers on the nature of the solvent suggests that the excited-state, non-radiative processes are affected by solvation. In polar, hydroxylic solvents, values of the fluorescence quantum yield for the non proton-transferred form are significantly lower, and the fluorescence lifetimes are shorter, than those calculated for aprotic solvents. This supports the proposal of the formation, in alcoholic solvents, of an excited-state encounter complex which facilitates ESIPT. The observed concentration dependence of the fluorescence lifetime and intensity of the blue emission from TIN in polymer films provides evidence for a non-radiative, self-quenching process, possibly due to aggregation of the stabilizer molecules.

The results obtained in this work indicate that both the structure of the stabilizer and the nature of the surrounding environment are important factors in determining the efficiency of the energy dissipation processes in these derivatives. Molecular structures in which the planar form is favoured and where the intra-molecular hydrogen bond is protected from interactions with the medium by the incorpoation of bulky substituent groups, should exhibit highest photostability and impart improved photoprotection to polymer substrates.

Acknowledgments

This work is supported by the Australian Research
Grants Scheme. Helpful discussions with Mr. I. H. Leaver
of the C.S.I.R.O. Division of Protein Chemistry are
gratefully acknowledged. The authors wish to thank
Mr. M. D. Yandell for his assistance with some of the
measurements reported in this work.

Literature Cited

1. Cicchetti, O. Adv. Polym. Sci. 1970, 7, 70.
2. Nicholls, C.H. In Developments in Polymer
   Photochemistry-1; Allen, N.S., Ed.; Applied Science
   Publishers: London, 1980; Chapter 5.
3. Hardy, W.B. In Developments in Polymer
   Photochemistry-3; Allen, N.S., Ed.; Applied Science
   Publishers: London, 1982; Chapter 8.
4. Heller, H.J. Eur. Polym. J. - Supp. 1969, 105.
5. Heller, H.J.; Blattman, H.R. Pure Appl. Chem. 1974,
   36, 141.
6. Barbara, P.F.; Rentzepis, P.M.; Brus, L.E. J. Am.
   Chem. Soc. 1980, 102, 2786.
7. Teruhiko, N.; Yamauchi, S.; Hirota, N.; Baba, M.;
   Hanazaki, I. J. Phys. Chem. 1986, 90, 5130.
8. Bigger, S.W.; Ghiggino, K.P.; Leaver, I.H.; Scully,
   A.D. J. Photochem. Photobiol.(A) 1987, 40, 391.
9. Otterstedt, J.A. J. Chem. Phys. 1973, 58, 5716.
10. Werner, T. J.Phys. Chem. 1979, 83, 320.
11. Huston, A.L.; Scott, G.W.; Gupta, A. J. Chem. Phys.
    1982, 76, 4978.
12. Flom, S.R.; Barbara, P.F. Chem. Phys. Lett. 1983, 94,
    488.
13. Bocian, D.F.; Huston, A.L.; Scott, G.W. J. Chem.
    Phys. 1983, 79, 5802.
14. Ghiggino, K.P.; Scully, A.D.; Leaver, I.H. J. Phys.
    Chem. 1986, 90, 5089.
15. Woessner, G.; Goeller, G.; Kollat, P.; Stezowski,
    J.J.; Hauser, M.; Klein, U.K.A.; Kramer, H.E.A. J.
    Phys. Chem. 1984, 88, 5544.
16. Woessner, G.; Goeller, G.; Rieker, J.; Hoier, H.;
    Stezowski, J.J.; Daltrozzo, E.; Neureiter, M.;
    Kramer, H.E.A. J. Phys. Chem. 1985, 89, 3629.
17. Ghiggino, K.P.; Scully, A.D.; Bigger, S.W.; Leaver,
    I.H. J. Polym. Sci., Polym. Chem. Ed. 1987, 25, 1619.
18. Huston, A.L.; Scott, G.W. J. Phys. Chem. 1987, 91,
    1408.
19. Lee, M.; Yardley, J.T.; Hochstrasser, R.M. J. Phys.
    Chem. 1987, 91, 4612.
20. Woolfe, G.J.; Thistlethwaite, P.J. J. Am. Chem. Soc.
    1981, 103, 6916.
21. Ghiggino, K.P.; Skilton, P.F.; Thistlethwaite, P.J.
    J. Photochem. 1985, 31, 111.

22. Melhuish, W.H. J. Phys. Chem. 1961, 65, 229.
23. Leaver, I.H. Aust. J. Chem. 1977, 30, 87.
24. Fleming, G.R.; Morris, J.M.; Robinson, G.W. Aust. J. Chem. 1977, 30, 233.
25. Werner, T.C. In Modern Fluorescence Spectroscopy; Wehry, E.L. Ed.; Plenum Press: New York, 1976; Chapter 7.
26. Lawton, W.H.; Sylvestre, E.A. Technometrics 1971, 13, 617.
27. Warner, I.M.; Christian, G.D.; Davidson, E.R.; Callis, J.B. Anal. Chem. 1977, 49, 564.
28. Aartsma, T.J.; Gouterman, M.; Jochum, C.; Dwiram, A.L.; Pepich, B.V.; Williams, L.D. J. Am. Chem. Soc. 1982, 104, 6278.
29. Davis, C.J.; Statistics and Data Analysis in Geology; Wiley and Sons: New York, 1973.
30. McMorrow, D.; Aartsma, T.J. Chem. Phys. Lett. 1986, 125, 581.
31. Shlyapintokh, V.Ya. Photochemical Conversion and Stabilization of Polymers; Hanser: Munich, 1984.
32. Reichardt, C. Solvent Effects In Organic Chemistry; Verlag-Chemie: New York, 1978.

RECEIVED July 13, 1988

Chapter 6

Radiolysis of Polycarboxylic Acids and Model Compounds

K. T. Campbell, D. J. T. Hill[1], James H. O'Donnell, P. J. Pomery, and C. L. Winzor

Polymer and Radiation Group, Department of Chemistry, University of Queensland, St. Lucia, Brisbane 4067, Australia

Gamma radiolysis of simple carboxylic acids and N-acetyl amino acids results in loss of the carboxyl group with formation of carbon monoxide and carbon dioxide. In the carboxylic acids, the ratio of CO/CO_2 produced is approximately 0.1, while in the N-acetyl amino acids the ratio is much smaller. In the poly carboxylic acids and poly amino acids, radiolysis also results in the loss of the carboxyl group, but here the ratio of CO/CO_2 is greater than 0.1. Incorporation of aromatic groups in the poly amino acids provides some protection for the carboxyl group. The degradation of the poly acids is believed to involve radical and excited state pathways.

Recently there has been increasing interest in studies of the effects of high energy radiation on polymers. Some of this interest has arisen because of the use of polymers as resists in the microchip industry, and some through the search for radiation resistant polymers for the aerospace and other high technology industries.

Studies of the physical and chemical changes which take place in polymers when they are exposed to high energy radiation have often drawn upon a knowledge of the corresponding changes which take place in suitable small molecule, model compounds. However, it must always be borne in mind that the radiation chemistry of small molecule models can be dominated by end-effects, which are not typical of a high molecular weight polymer.

In this paper we describe some of our work on a study of the effects of gamma radiation on a variety of polycarboxylic acids and a related series of small molecule, model compounds containing carboxyl groups.

[1]Address correspondence to this author.

0097–6156/89/0381–0080$06.00/0
© 1989 American Chemical Society

RADIATION EFFECTS IN POLYMERS

The primary event which takes place when high energy radiation, such as gamma radiation, interacts with a polymer molecule involves the ejection of an electron, with formation of the polymer cation radical, as shown in Equation (1):

$$P \xrightarrow{\gamma} P^{\overset{+}{\bullet}} + e^- \qquad (1)$$

The ejected electron may become trapped on a suitable site in the matrix of the polymer, thus forming a polymer anion radical.

$$P + e^- \longrightarrow P^{\overset{-}{\bullet}} \qquad (2)$$

Alternatively, the electron, or the polymer anion, may react with an existing cation radical producing an excited state of the polymer molecule, P*. For example,

$$P^{\overset{+}{\bullet}} + e^- \longrightarrow P*$$

Polymer cation radicals, anion radicals and excited state species are all very reactive, so that further chemistry will generally take place. Polymer cation radicals are usually reactive even at temperatures below 77K, and often decompose to produce a polymer radical and a cation, which is often H^+. Polymer anion radicals are usually less reactive than the cation radicals, and are often stable at 77K, but they are usually unstable at room temperature. Excited state species can undergo decomposition by a variety of routes including: (i) homolytic cleavage to form two neutral radicals, (ii) heterolytic cleavage to form an anion and a cation, or (iii) bond rupture with the formation of two neutral molecules.
    The neutral polymer radicals which are produced also often undergo further reactions, which can result in chemical changes in the polymer. These reactions may include crosslinking or scission of polymer chains, formation of small molecule products, changes in the stereochemistry of the polymer chains, changes in the crystallinity of the polymer or a variety of other chemical and physical processes.
    The radiation chemistry of the polycarboxylic acids (and the model compounds) is highlighted by many of these reactions.

A. MODEL COMPOUNDS

ALIPHATIC CARBOXYLIC ACIDS The radiation chemistry of the simple aliphatic carboxylic acids has been widely investigated. The major products of gamma radiolysis of these compounds are typified by those found (1) for radiolysis of isobutyric acid at 273 K (Table I).

Table I. Yields of volatile products for gamma radiolysis
of isobutyric acid at 273 K

Product	G-value
propane	3.72
propene	1.00
methane	0.10
hydrogen	0.87
carbon dioxide	5.44
carbon monoxide	0.41
water	1.00

The major hydrocarbon product is the parent alkane, propane, formed by loss of the carboxyl group. Smaller amounts of other fragments, such as propene, methane and hydrogen are also observed. These are most likely formed as a result of reactions of the propane radical; for example propene can be formed by disproportionation of the propane radical:

$$CH_3-\overset{\bullet}{C}H-CH_3 + CH_3-\overset{\bullet}{C}H-CH_3 \longrightarrow CH_2=CH-CH_3 + CH_3-CH_2-CH_3$$

Carbon dioxide is the major product of the radiolysis, and results from the loss of the carboxyl group. Carbon monoxide is found in somewhat smaller yield than carbon dioxide. Water is also produced, but its yield is often difficult to quantify because of the strong hydrogen bonding which exists between water and carboxylic acid groups. Hydrogen is not normally found in high yield.

For a wide variety of aliphatic carboxylic acids, including some with aromatic substituents, the product yields for CO and CO_2 are in an approximately constant ratio of 1:10 for radiolysis at 298 K. Values for some typical compounds are given below in Table II.

The major radical intermediates formed following radiolysis of aliphatic carboxylic acids are also typified by those found following radiolysis of isobutyric acid (1). In Figure 1 are illustrated the ESR spectra found following radiolysis of this acid at 77 K and 195 K.

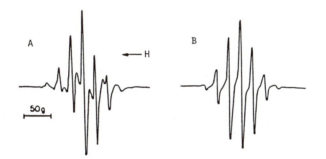

Figure 1. ESR spectra of isobutyric acid following gamma radiolysis in the solid state at (A) 77 K (B) 195 K.

Table II. G values for CO and CO_2 for γ radiolysis
of carboxylic acids at 298 K

Acid	G(CO)	$G(CO_2)$	$G(CO)/G(CO_2)$
CH_3COOH[a]	0.73	6.0	0.12
CH_3CH_2COOH[a]	0.46	4.4	0.10
$(CH_3)_2CHCOOH$[b]	0.41	5.4	0.08
ϕCH_2COOH[c]	0.37	3.8	0.10
$\phi(CH_2)_2COOH$[c]	0.20	3.0	0.06

a   reference 2
b   reference 1
c   Klika, K; Hill, D.J.T.   unpublished results

The ESR spectrum of isobutyric acid following radiolysis and measurement at 77 K comprises several components. These have been identified as: (i) the decarboxylation radical (I), which appears as an octet with a splitting of 23.5 gauss (ii) the anion radical (II), which appears as a doublet with a splitting of 24 gauss and (iii) the hydrogen abstraction radical (III), the major component, which appears as a septet with a splitting of 21.4 gauss.

Radiolysis of isobutyric acid at 195 K results in the formation of only one radical intermediate, the hydrogen abstraction radical III. The decarboxylation radical and the anion radical are both unstable at this temperature and react forming the abstraction radical and other products. The hydrogen which is abstracted is generally that which is attached to the carbon atom α to the carboxyl group.

Since decarboxylation is a primary reaction of the aliphatic carboxylic acids, it is interesting to compare the radical yield measured at 77 K (3) with the yield of carbon monoxide plus carbon dioxide. These values are compared in Table III. The results suggest that there is a correlation between the loss of the carboxyl group and the formation of radicals in the carboxylic acids.

Table III. $G(R\cdot)$ at 77 K compared to $G(CO) + G(CO_2)$
at 298 K for a series of carboxylic acids

Acid	$G(R\cdot)$	$G(CO) + G(CO_2)$
CH_3COOH	4.9	4.4
CH_3CH_2COOH	6.7	4.9
$CH_3(CH_2)_2COOH$	5.4	5.0
$(CH_3)_2CHCOOH$	5.7	5.9
$(CH_3)_3CCOOH$	5.2	6.3[a]

[a] determined at $80°C$

Based upon the observed radical and molecular product, it has been suggested by ourselves and others (3,4) that the major mechanism for the degradation of simple carboxylic acids is:

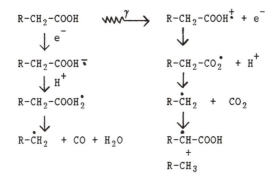

N-ACETYL AMINO ACIDS The N-acetylamino acids have been suggested as model compounds for study of the degradation of polyamino acids and proteins. However, they contain a free carboxyl group, so might also be representative of those polyamino acids which contain carboxyl groups in their side chain. The results of our investigation (5) of the volatile products which are formed on gamma radiolysis of N-acetyl glycine at 303 K are given in Table IV. These products are typical of the observations for other, similar compounds.

The range of products formed on gamma radiolysis of N-acetylglycine was similar to that formed on radiolysis of the aliphatic carboxylic acids, but there are some noticeable differences in the yields of products. Carbon dioxide is by far the major volatile product of radiolysis and the corresponding product of the decarboxylation reaction, N-methyl acetamide, is also present in large yield, but the yield of this product was not quantitatively determined. By contrast, carbon monoxide is found in very small yield. The yield of acetamide, the product of N-Cα bond scission, is found in much greater yield.

These observations are common to the series of N-acetylamino acids which have been investigated. Whereas the carbon monoxide to carbon dioxide yield for the aliphatic carboxylic acids are in the approximate ratio of 1:10, the corresponding ratio for the N-acetyl amino acids is much smaller, because of the much smaller yields for carbon monoxide. This suggests that there must be another site on which the ejected electron can be trapped with greater efficiency than that for trapping on the carboxyl group.

The yields of carbon monoxide and carbon dioxide are given in Table V for a series of amino acids with aliphatic or aromatic side chains. In all cases, the ratio of G(CO) to G(CO$_2$) is much less than that reported in Table II for the corresponding carboxylic acids.

A variety of radical products is observed following gamma radiolysis of the N-acetyl amino acids at 77 K (6), depending on the nature of the side chain of the parent amino acid. In the case of N-acetyl alanine, for example, the intermediates are: (i) the anion radical IV (ii) the decarboxylation radical V (iii) the deamination radical VI and (iv) the alpha carbon radical VII.

Table IV. G-values for the volatile products formed on gamma radiolysis of N-acetylglycine at 303 K

product	G-value
CO_2	2.0
CO	0.004
H_2	0.22
CH_4	0.007
CH_3CONH_2	0.09
CH_3COOH	0.05
$CH_3CONHCH_3$	present

Table V. G-values for CO and CO$_2$ for γ-radiolysis of N-acetyl amino acids, CH_3-CO-NH-CH(R)-COOH, at 303 K

Amino acid	G(CO)	G(CO$_2$)	G(CO)/G(CO)$_2$
R = H	0.004	2.0	0.002
R = CH_3	0.02	4.8	0.004
R = $CH(CH_3)_2$	0.03	8.2	0.004
R = ϕ	0.06	1.2	0.05
R = $CH_2\phi$	0.02	1.9	0.01

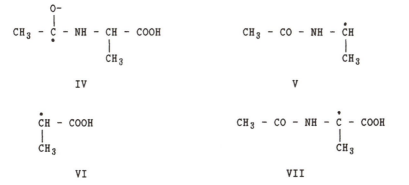

As the temperature is raised above 77 K, the anion radical IV disappears at approximately 190 K and at room temperature only the carbon centred radicals V - VII are observed.

A similar behaviour has been found to occur with the other N-acetyl amino acids. In each case, the most stable radical observed at 303 K was the alpha carbon radical, as was also observed for the aliphatic carboxylic acids. In Table VI the radical yields observed following gamma radiolysis of a series of N-acetyl amino acids at 303 K are reported, together with the stable radical intermediates observed at this temperature (5).

Table VI. Observed radicals and measured radical yields for gamma radiolysis of the N-acetyl amino acids, CH_3-CO-NH-CH(R)-COOH, at 303 K

Amino acid	Radical Assignments	G(R·)
R = -H	CH_3-CO-NH-$\overset{\bullet}{C}$H-COOH	2.8
R = -CH_3	CH_3-CO-NH-$\overset{\bullet}{C}$(CH_3)-COOH	3.7
	CH_3-CO-NH-$\overset{\bullet}{C}$H(CH_3)	
	$\overset{\bullet}{C}$H(CH_3)-COOH	
R = -CH(CH_3)$_2$	CH_3-CO-NH-$\overset{\bullet}{C}$(CH (CH_3)$_2$)COOH	6.8
	CH_3-CO-NH-CH($\overset{\bullet}{C}$ (CH_3)$_2$COOH	
R = -ϕ	not assigned	1.6
R = -CH_2-ϕ	not assigned	1.8

Again the close correspondence between the measured radical
and carbon dioxide yields for γ-radiolysis of the N-acetyl amino
acids in the solid state suggests that the mechanisms for radical
production and carbon dioxide formation are closely related, as
they were for the aliphatic carboxylic acids. The following
mechanism has been proposed (5) in order to account for the major
degradation products and observed radical intermediates.

$$CH_3-CO-NH-CH(R)-COOH \xrightarrow{\gamma} CH_3-CO-NH-CH(R)-COOH^{+\cdot} + e^-$$

$$\downarrow e^-$$

$$(CH_3-CO-NH-CH(R)-COOH)^{-\cdot}$$

$$\downarrow$$

$$CH_3-CO-NH_2$$
$$+$$
$$R-CH_2-COOH$$
$$+$$
$$CH_3-CO-NH-CH_2-R$$
$$+$$
$$CO + H_2O$$

$$CH_3-CO-NH-\overset{\bullet}{C}H(R) + CO_2 + H^+$$

$$\downarrow$$

$$CH_3-CO-NH-\overset{\bullet}{C}(R)-COOH$$
$$+$$
$$CH_3-CO-NH-CH_2-R$$

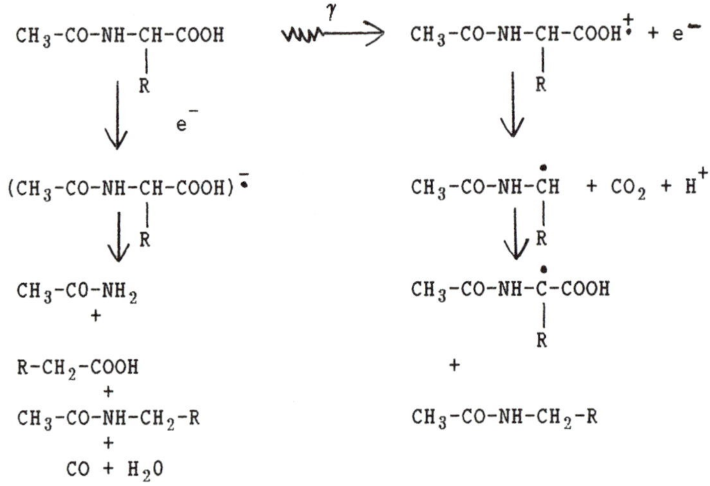

B. POLY ACIDS

POLYCARBOXYLIC ACIDS The gamma radiolysis of the homopolymers of
acrylic, methacrylic and itaconic acids have been investigated in
the solid state at 303 K, and in each case the yields of carbon
monoxide, carbon dioxide and of radical intermediates have been
measured. These are reported in Tables VII and VIII respectively.

Table VII. G-values for carbon monoxide and carbon dioxide
formed on radiolysis of poly acids at 303 K

Poly acid	$G(CO)$	$G(CO_2)$	$G(CO)/G(CO_2)$
Acrylic	3.1	9.0	0.34
Methacrylic	2.8	7.9	0.35
Itaconic	1.8	3.6	0.50

Table VIII.   G-values for radical formation for gamma
radiolysis of poly acids at 303 K

Acid	Radical intermediates	G(R•)
Acrylic	$\sim CH_2-\overset{\bullet}{C}-CH_2\sim$ (90%) $\qquad\quad$ COOH $\sim CH-\overset{\bullet}{C}H-CH\sim$ (10%) $\quad$ COOH $\quad$ COOH	3.8
Methacrylic	$\qquad\quad CH_3$ $\qquad\quad\mid$ $\sim CH_2-\overset{\mid}{\underset{\mid}{C}}\bullet$ $\qquad\quad COOH$	3.1
Itaconic	not assigned	3.1

As with the aliphatic carboxylic acid model compounds, the
major volatile product observed on gamma radiolysis of the poly
acids is carbon dioxide. However, the carbon dioxide yields are
somewhat larger than those observed for the model compounds.
Carbon monoxide was also observed in significant yields for each
of the poly acids and the ratio of $G(CO)/G(CO_2)$ was found to be
approximately three times greater than that found for the small
molecule model compounds. This suggests that the processes
involved in the formation of carbon monoxide and carbon dioxide
are both more efficient in the poly acids.

While the major stable radical intermediate for polyacrylic
acid was the alpha carbon radical, as expected on the basis of the
model compound studies, a small amount (ca. 10 per cent) of the
radical formed by abstraction from the methylene carbon was also
observed.

The only radical intermediate observed for poly methacrylic
acid was the propagating radical formed by main chain scission.
This observation is similar to that noted for gamma radiolysis of
poly methylmethacrylate, where the propagating radical is also
found as the only stable radical intermediate following radiolysis
at 303 K. In both cases the propagating radical is formed by
β-scission following the loss of the side chain, resulting in
formation of the unstable tertiary radical.

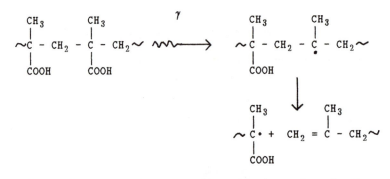

For each of the poly carboxylic acids investigated, the sum of the yields of carbon monoxide and carbon dioxide is much larger than the yield of radical products. This observation differs from that noted for the model compounds, where the two were of similar magnitude. This suggests that excited state processes may play a more significant role in the degradation of the poly acids than they do in the small molecule, model compounds.

Thus, polymethacrylic acid undergoes net scission on gamma radiolysis with a G-value of approximately 4, while polyacrylic acid, on the other hand, undergoes net crosslinking with a G-value of approximately 1.2 (7). Crosslinking in polyacrylic acid is favourable because of the formation of main chain radicals. These can react to form crosslinks between polymer chains.

Incorporation of aromatic groups into a copolymer, such as poly(methyl methacrylate-co-styrene) has been reported (8) to result in protection of methyl methacrylate groups against radiation damage. This can be observed through the G-values for production of small molecule volatile products or the G-values for production of radical intermediates. However, gamma radiolysis of poly(methacrylic acid-co-styrene) copolymers do not show a similar protective effect for radical production, as demonstrated in Figure 2. Here there is an approximately linear dependence of radical yield on polymer composition, indicating that there is no protection against radical formation in these copolymers. This suggests that the mechanism for radical production in this copolymer (that is, decarboxylation of the polymer chain) is a highly efficient process, and competes effectively with energy transfer to the aromatic group.

POLYAMINO ACIDS Aliphatic polyamino acids irradiated in the solid state have been reported to undergo N-Cα, main-chain, bond scission on gamma radiolysis (9) and the stable radical intermediates formed following radiolysis at 303 K are alpha carbon radicals, as observed in the N-acetyl amino acids. However, the major reaction following radiolysis of poly glutamic acid is decarboxylation (Hill, D.J.T.; Ho, S.Y.; O'Donnell, J.H.; Pomery, P.J. Radiat. Phys. Chem., submitted for publication),

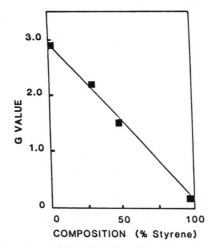

Figure 2.   G-value for radical production for poly(methacrylic acid-co-styrene) as a function of copolymer composition.

with formation of carbon dioxide, $G(CO_2) = 5.5$, and carbon monoxide, $G(CO) = 2.2$, as was observed for the polycarboxylic acids.

The carbon monoxide yield is much larger than that expected on the basis of the carboxylic acid model compounds, and is similar to that observed for the polycarboxylic acids, with $G(CO)/G(CO_2) = 0.4$.

Two stable radical intermediates are observed following gamma radiolysis at 303 K. The alpha carbon radical VIII and the side chain radical IX are formed in approximately equal yields, with the total G-value for radical production equal to 3.2. This value is similar to that observed for the poly acids. The observed radicals are those which would be expected on the basis of the aliphatic carboxylic acids and previous studies of the poly amino acids with aliphatic side chains.

$$\sim NH - \overset{\bullet}{C} - CO \sim \qquad\qquad \sim NH - CH - CO \sim$$
$$| \qquad\qquad\qquad\qquad\qquad |$$
$$CH_2 \qquad\qquad\qquad\qquad CH_2$$
$$| \qquad\qquad\qquad\qquad\qquad |$$
$$CH_2 \qquad\qquad\qquad\qquad \bullet CH$$
$$| \qquad\qquad\qquad\qquad\qquad |$$
$$COOH \qquad\qquad\qquad\qquad COOH$$

$$\text{VIII} \qquad\qquad\qquad\qquad\qquad \text{IX}$$

The effect of aromatic groups on the radiation chemistry of glutamic acid residues in a polymer has been studied by investigation of a series of random copolymers of glutamic acids and tyrosine. The results of this investigation are shown in Figure 3 for the variation in the yields of carbon dioxide, carbon monoxide and radical intermediates with copolymer composition. By contrast with the results observed for poly(methacrylic acid-co-styrene), poly(glutamic acid-co-tyrosine) shows a distinct protective effect due to the presence of the aromatic group in the tyrosine side-chain.

CONCLUSIONS

The carboxyl group in aliphatic carboxylic acids is very sensitive to gamma radiation and undergoes degradation by both anionic and cationic pathways with the formation of carbon monoxide and carbon dioxide respectively. The degradation of the carboxyl group in N-acetyl amino acids results in a much smaller yield of carbon monoxide than for the aliphatic carboxylic acids. This is believed to be a result of much more efficient trapping of the ejected electron by the peptide group in these compounds.

In the poly carboxylic acids, carbon dioxide is the major product of radiolysis, but the carbon monoxide yields are greater than they are for the aliphatic carboxylic acids. However, the radical yields are not greater than expected on the basis of the model compounds, which suggests that excited states play an important role in the degradation of these poly acids.

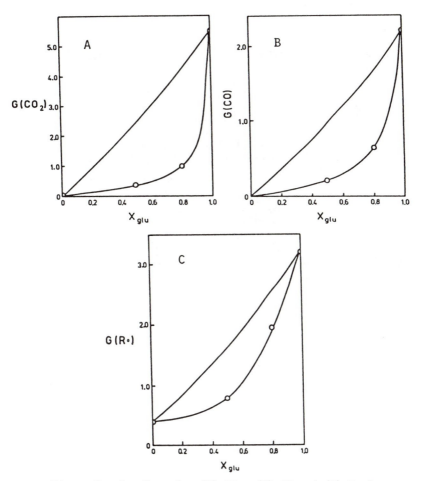

Figure 3. G-values for (A) CO_2 (B) CO and (C) R• for gamma radiolysis of poly(glutamic acid-co-tyrosine) at 303 K.

Incorporation of aromatic residues into the side chain of a poly carboxylic acid does not provide a protection against radical formation, indicating that decarboxylation is more efficient than energy transfer in these polymers. However, incorporation of tyrosine residues into a polyglutamic acid copolymer does result in efficient protection against radiation damage, indicating that energy transfer is more efficient in these copolymers.

LITERATURE CITED

1. O'Donnell, J.H.; Sothman, R.D. Radiat. Phys. Chem. 1979, 13, 77.
2. Johnsen, G.H. JACS 1960, 63, 2041.
3. Mach, K.; Janovsky, I.; Vacek, K. Coll. Czech. Chem. Commun. 1979, 44, 3632.
4. Ayscough, P.B.; Oversby, J.P. Trans. Faraday Soc. 1971, 67, 1365.
5. Hill, D.J.T.; Ho, S.Y.; Garrett, R.W.; O'Donnell, J.H.; O'Sullivan, P.W.; Pomery, P.J. Radiat. Phys. Chem. 1981, 17, 163.
6. Hill, D.J.T.; Ho, S.Y.; O'Donnell, J.H.; Pomery, P.J. Radiat. Phys. Chem. 1985, 26, 191.
7. Lawler, J.P.; Charlesby, A. Europ. Polym. J. 1975, 11, 755.
8. Busfield, W.K.; O'Donnell, J.H.; Smith, C.A. Polymer 1982, 23, 431.
9. Hill, D.J.T.; Ho, S.Y.; O'Donnell, J.H.; O'Sullivan, P.W.; Pomery, P.J. Polym. Degrad. and Stability 1980, 3, 83.

RECEIVED September 1, 1988

Chapter 7

Photophysical Studies of Spin-Cast Polymer Films

L. L. Kosbar[1], S. W. J. Kuan[2], C. W. Frank[2,4], and R. F. W. Pease[3]

[1]Department of Chemistry, Stanford University, Stanford, CA 94305
[2]Department of Chemical Engineering, Stanford University, Stanford, CA 94305
[3]Department of Electrical Engineering, Stanford University, Stanford, CA 94305

Spin casting is a commonly used technique for producing thin, uniform polymer films, especially for semi-conductor fabrication. There has been little attention given to the response of the polymers at the molecular level, however. We have reviewed previous literature on the spin casting process, as well as pertinent aspects of solvent cast and oriented polymer films. Our experimental work has focused on the use of excimer fluorescence to elucidate polymer chain conformation and fluorescent probe environment. The effects of spin speed and polymer molecular weight on spin cast polystyrene films and subsequent annealing behavior have been studied. In addition, the aggregation of a small molecule dye (pyrene) in novolac films was investigated.

The U.S. - Australia Symposium on Radiation Effects on Polymeric Materials contained research presentations on fundamental radiation chemistry and physics as well as on technological applications of polymer irradiation. This paper represents a hybrid contribution of these two areas, examining a field of extensive technological importance. Spin casting of radiation sensitive polymer resists for microelectronic fabrication was studied using photophysical techniques that are sensitive to the fundamental radiation response in the ultraviolet range.

Thin, uniform polymer films can be formed by casting a polymer solution onto a rotating disc. Investigations of spin casting have, to date, focused on the process at a macroscopic level. Of primary concern has been the ability to predict the final film thickness and any radial dependence of the thickness induced by spinning. Various models have been derived to relate the spinning and solution characteristics to film thickness and uniformity. These parameters are of very practical importance in the present applications of spin casting, especially for resists.

[4]Address correspondence to this author.

0097–6156/89/0381–0095$06.00/0
© 1989 American Chemical Society

The intent of our investigations is to study the effects of
spincasting on polymers at a molecular level. The polymers may be
radiation sensitive materials such as poly(methyl methacrylate)
(PMMA), polysulfones, or polystyrene derivatives, or radiation
insensitive materials such as novolac or pure polystyrene. In all
cases, photophysical techniques will be used to interrogate the
polymer structure and film homogeneity (for the films which are
mixtures of a radiation insensitive polymer and a radiation sensitive material). There has been little effort given to understanding
the effects of spin casting on configuration of polymer chains up
until now. Indeed, for circuit geometries above the sub-micron
regime, it is less important because any small inconsistencies or
inhomogeneities in the resist film are within the allowable tolerances. But as image dimensions creep down into the 0.5 μm range,
and tolerances are reduced into the realm of polymer dimensions, the
behavior of the film on the molecular level may play an important
role in enhancing resist control and reproducibility.

Spin casting applies radial stress to the polymer film as the
solvent evaporates. As a result, it is likely that the polymer
chains will be frozen into stretched, non-equilibrium configura-
tions. Radial orientation and stress in the polymer backbone could
slightly change the response of the polymer to radiation or solvent
developers. Release of the stress may appear in the form of microcracking (1) of the film, or loss of adhesion with the substrate
(2). Resist films generally receive a post spinning bake at a
temperature which may be above or below the glass transition temperature (Tg) of the polymer, depending on the resist. Many resists,
such as most positive diazoquinone resists are too thermally sensitive to be baked above the Tg of the polymer. The bake is intended
to remove any remaining trapped solvent and to relax stresses
induced by spinning, but there has been no thorough study of the
effectiveness of commonly used baking conditions. Aggregation of
individual components of the resist during spinning could change the
development characteristics of the resist, the effect of which would
be dependent on the size of the aggregated domains. In particular,
some roughening of the image sidewalls or overall dimensional variation would be expected.

Intrinsic and extrinsic fluorescence can be used as very
sensitive probes of the polymer environment and configuration. A
particular example is excimer fluorescence, which can occur when two
aromatic groups interact in a coplanar structure. The rings must be
within three to four angstroms to produce a suitable excimer forming
site, so the excimer fluorescence yields information on the local
concentration of these properly oriented aromatic chromophores. Excimer formation may be intermolecular if the two aromatic rings are
on different polymer chains, or intramolecular if they are on repeat
units of the same polymer chain. Changes in the ratio of excimer to
monomer fluorescence intensity (Ie/Im) of aromatic polymers such as
polystyrene have been used to measure the level of orientation in a
uniaxially stretched film (3). This technique can be applied to the
radial orientation caused by spin casting. The fluorescence of
probes such as pyrene can also yield a variety of insights into the
effects of spin casting. Probes that are chemically attached to the
polymer chain can yield configurational information on otherwise
non-fluorescent polymers. Aggregation of free fluorescent probes

can be monitored and used as a model system for polymer/dye type
positive resist systems. We will present the initial results obtained using these fluorescent techniques to evaluate the extent of
orientation and dye aggregation induced by spin casting and the
effectiveness of annealing to remove orientation and stress in
polymer systems.

Background on Spin Casting. As early as 1958, Emslie, et al. (4)
proposed a theoretical treatment of spin casting for nonvolatile
Newtonian fluids. This theory predicted that films formed on a flat
rotating disc would have radial thickness uniformity. They predicted that the final film thickness would depend on spin speed (ω)
and viscosity (η) as well as other variables such as liquid density
and initial film thickness. The dependence of thickness on ω and η
was also recognized by many of the other authors reviewed in this
paper, and their proposed relationships are compared in Table I.
Acrivos, et al. (5) extended the Emslie treatment to the general
case of non-Newtonian fluids, a category into which most polymers
fall. Acrivos predicted that non-Newtonian fluids would yield films
with non-uniform radial thickness.

Meyerhofer (6) included the effects of solvent evaporation
during spinning in his model, which he found to be in good agreement
with experimental results for photoresist films. Chen (7) also recognized the importance of solvent evaporation on final film thickness. Chen pointed out that as the solvent evaporates, it will cool
the solution, causing an increase in viscosity. He claimed that the
film thickness should be related to the physical parameters of the
solvent such that the same thickness film could be attained with a
low viscosity solution of a highly volatile solvent or with a high
viscosity solution of a low-volatility solvent. Weill (8) noted
that the viscosity of a polymer solution is dependent on both the
polymer concentration and molecular weight (MW). Weill found that
the thickness of films cast from several solutions with the same
viscosity but varying MW increased with MW, although no exact
relation between MW and final thickness was derived.

Jenekhe (9,10) also incorporated the changing rheological properties of the drying film into his thickness model. He defined an
experimentally determined parameter (α) that measures the time-
dependent change in viscosity due to solvent evaporation during
casting, with $\alpha=0$ for nonvolatile fluids. A range of $\alpha=0.44$-3.00
was found to account for previous empirical results for spin cast
films. Jenekhe found that many polymer solutions, including those
of interest to the microelectronics industry, can be described by
the Carreau non-Newtonian viscosity equation, which was shown to
predict radially uniform films at sufficiently long spinning times.
Flack, et al. (11) took the changing rheological properties into
account by allowing viscosity to vary with concentration and shear
rate during spinning. They determined that convective radial flow
dominated the calculated thickness during the early stages, and that
solvent evaporation became more important in the later stages of the
spinning process. They also indicated that a radial thickness
dependence should be expected if a ramp of the spin speed were used.
Bornside, et al. (12) are also attempting to separate the spinning
process into deposition, spin up, spin off, and evaporation stages,
and to model these stages using fewer of the simplifying assumptions

employed by previous authors. The model promises to be very complex.

Several authors have tried to determine empirical relations that will accurately predict the dependence of final film thickness on ω, η, and other parameters. The relationships obtained by Damon (13), Daughton and Givens (14), and Malangone and Needham (15) are included in Table I for comparison with the predictions of the theoretical models.

Table I. The Predicted Dependence of Film Thickness (h) on ω and η according to $h \propto \omega^n \eta^m$

Reference	n	m	Other Variables
Emslie (4)	-1	0.5	density, spinning time, initial thickness
Damon (13)	-0.5	-	initial concentration
Meyerhofer (6)	-0.67	0.33	evaporation rate, initial concentration
Daughton (14)	-0.5-0.8	0.29	spinning time, initial concentration
Malangone (15)	-0.5	0.33	initial concentration
Chen (7)	-0.5	0.36	evaporation rate, heat capacity, latent heat of evaporation
Jenekhe (9)	$-2/(2+\alpha)$	$1/(2+\alpha)$	spinning time, initial thickness, index of viscosity change (α)

It is obvious from the previous discussion and Table I that there has been no general agreement on the effect of ω and η on thickness, and even less agreement on what other variables are significant in determining final film thickness. While the exponent for viscosity tends to be about 0.3 in most of the models, the empirical exponent of spin speed varies from -0.5 to -0.8. The various models also give different relative weighting to the two parameters. The variety of other variables that were included in the models indicate that there is still a great deal of disagreement as to what the critical parameters for spin casting are. The macroscopic properties of spin cast films are difficult to predict from first principles, and do not lend themselves to empirical analysis either. Daughton and Givens (14) found different empirical relations to apply for different polymer systems. These simple models are giving way to much more complicated, multi-step models, but there is obviously still much that is not understood about the "simple" macroscopic aspects of spin casting.

Resist systems may be more complicated than just a single polymer in a single solvent. They may be composed of polymer, polymer/dye, or polymer/polymer combinations (where the small molecule dye or additional polymer increases the radiation sensitivity of the resist film) with one or more solvents. The more complicated polymer/dye or polymer/polymer systems have the added possibilities of phase separation or aggregation during the non-equilibrium casting process. Law (16) investigated the effects of spin casting on a

small molecule dye in solution with polymers. He found a slight
increase in the aggregation of the dye at high dye concentration
(>40%) and low spin speed. Law suggested that spinning conditions
could affect the kinetically controlled aggregation of polymer/dye
systems with poor compatibility.

Even at their best, the models are able to predict only macroscopic properties of the films, yielding no information on microscopic parameters that may affect resist performance. It is highly
probable that spin casting induces some structure or preferential
chain orientation into the films, or causes secondary effects such
as the aggregation observed by Law. These effects are barely addressed in the currently available literature. However, some
earlier works (3,17-19) on solvent (static) cast films have investigated the molecular orientation of polymer chains as well as chain
relaxation due to thermal annealing.

Prest and Luca (17) found that the solvent-casting process
preferentially aligned polymeric chains in the plane of the film, as
measured by the optical anisotropy of the films. Cohen and Reich
(18) reported that films cast from monodisperse low molecular weight
(< 30,000) polystyrene exhibit high ordering close to the substrate,
but this ordering decays within 1 μm from the substrate. Films cast
from monodisperse high molecular weight polystyrene (> 30,000), on
the other hand, exhibit a much smaller, but very long-range (~10 μm)
degree of order. Croll (19) studied the formation of stresses in
solvent cast films as the solvent evaporates. The thickness of the
film will decrease, but the area is constrained by adhesion to the
substrate. Because of this constraint, internal stress arises in
the plane of the coating. He further concluded that the residual
internal stress is independent of dried coating thickness and initial solution concentration. Polymer orientation has also been
studied with stretched polymer films. Jasse and Koenig (20) studied
uniaxially oriented polystyrene films using Fourier transform infrared spectroscopy and concluded that the orientation process produces
alignment of the chains as well as an increase in the amount of
trans conformational segments. We believe that the photophysical
approach described in this paper will allow us to investigate similar microscopic properties of spin cast films.

EXPERIMENTAL

Materials. Polystyrene samples of molecular weights 10,000, 50,000,
300,000 and 600,000 were obtained from Polysciences Inc.; they all
have narrow molecular weight distributions, with Mw/Mn < 1.1. They
were purified by three precipitations from tetrahydrofuran (THF)
into methanol. The novolac sample was provided by Kodak, and was
synthesized from pure meta-cresol and formaldehyde. The material
was quite polydisperse with Mw= 13,000 and Mw/Mn= 8.5, as measured
by gel permeation chromatography (GPC) with polystyrene standards.
The material was purified by three precipitations from THF into
hexane. The pyrene and pyrene butyric acid (PBA) were recrystallized from toluene prior to use. The THF was dried by refluxing and
distilling over sodium and benzophenone. Spectral grade dioxane
(Aldrich), THF (Alltech), 2-methoxyethyl ether (Aldrich), and
reagent grade methanol (Aldrich) were used.

Pyrene Tagging. The tagged novolac materials were prepared by the
reaction of pyrene butyric acid chloride (PBA-Cl) with the polymer.
This reaction covalently bonds the pyrene butyric acid to the novolac by forming an ester linkage through the phenolic group on the
novolac. The PBA-Cl was synthesized by reacting pyrene butyric acid
(3.42 g, 11.8 mmol) with oxalyl chloride (17.5 ml, 20.0 mmol) in
dichloromethane (75 ml) at room temperature, and stirred overnight.
The solvent and excess oxalyl chloride were pulled off under vacuum.
The crude product was dried under vacuum for two hours. To obtain
the most highly tagged sample, the crude PBA-Cl (2.0 g, ~6.5 mmol)
was dissolved in dry THF (15 ml) and added slowly to a stirred solution of novolac (2.1 g, 17.7 mmol monomer) and pyridine (10 ml), as
a base, in dry THF (50 ml). The reaction was performed under nitrogen. The mixture was refluxed for three hours, and then stirred at
room temperature overnight. The solution was washed twice with 10%
aqueous HCl and twice with water. The organic layer was dried over
MgSO$_4$, the polymer was precipitated three times from THF/hexane, and
the product dried in vacuum for 24 hours. Four samples tagged with
lower levels of pyrene were prepared in a similar manner using 1.2 g
of novolac with 0.2, 0.12, 0.05, or 0.02 g of PBA-Cl and 0.5-1 ml of
N,N-diisopropylethylamine, which is a better base than pyridine.
About 64-75% of the polymer was recovered after the reactions.
     To ensure that the pyrene was attached to the polymer chain, a
sample of the purified polymer was analyzed by GPC with tandem UV
and refractive index detectors. There was no detectable (<1%) free
pyrene in the sample. The MW and dispersity of the novolac also appeared to be unchanged by the reaction. The concentration of pyrene
tags was determined by UV absorption spectroscopy using solutions of
methyl-1-pyrene butyrate for calibration. It was found that the
polymers had 19.6, 5.03, 2.74, 1.18, and 0.43% of the monomer units
tagged with pyrene. Assuming a weight average molecular weight of
13,000, this would result in 21, 5, 3, 1.3 and 0.5 pyrene groups per
chain, respectively. It is assumed that the pyrene is randomly
attached along the polymer chain, although this has not been
experimentally verified.

Samples for Spectroscopic Analysis. Solvent cast films of PS (10 μm
thick) were prepared by casting from dioxane solution onto quartz
wafers at room temperature in a dry nitrogen ambient. Spin cast
films of PS (0.1 - 1.0 μm thick) were prepared by flooding the wafer
with a solution of PS in dioxane and spinning them in a dry nitrogen
ambient at a given spin speed for 60 seconds using a Metron Systems
Inc. LS-8000 spinner. The films were prepared under a dry nitrogen
ambient to reduce the uptake of moisture by the dioxane, which can
lead to poor quality films. In order to determine the effect of
residual solvent on fluorescence spectra, spectra of samples kept
under vacuum at room temperature for eight hours were compared with
spectra of samples which were not placed under vacuum. No differences were observed. Solutions of the novolac samples were prepared
with 2-methoxyethyl ether (diglyme) and were cast onto glass discs
using the LS-8000 spinner. Fluorescence measurements of novolac
solutions were made from solutions prepared with diglyme. The solutions were prepared to be 1 x 10^{-6} M pyrene, with the concentration
of polymer varying according to its tagging level.

Fluorescence Instrumentation and Measurements. Fluorescence spectra
of the PS samples were obtained on a steady state spectrofluorometer
of modular construction with a 1000 W xenon arc lamp and tandem
quarter meter excitation monochromator and quarter meter analysis
monochromator. The diffraction gratings in the excitation mono-
chromators have blaze angles that allow maximum light transmission
at a wavelength of 240 nm. Uncorrected spectra were taken under
front-face illumination with exciting light at 260 nm. Monomer
fluorescence was measured at 280 nm and excimer fluorescence was
measured at 330 nm, where there is no overlap of excimer and monomer
bands.
    Fluorescence spectra of the novolac samples were measured on a
Spex Fluorolog 212 spectrofluorometer with a 450 W xenon arc lamp
and a Spex DM1B data station. Spectra were taken with front-face
illumination using a 343 or 348 nm excitation wavelength for solutions or films, respectively, which are near the maximum transmission region of this spectrometer. Spectra were corrected using a
rhodamine B reference. Monomer fluorescence was measured at 374 or
378 nm and excimer fluorescence was measured at 470 nm. Monomer and
excimer peak heights were used in calculations of Ie/Im. The I_1
monomer peak of pyrene was used to reduce overlap with the excimer
emission.

RESULTS

Fluorescence of Polystyrene Films. Films of various MW polystyrene
were prepared at spin speeds from 1,000 to 8,000 rpm. The ratio of
excimer to monomer emission intensity, Ie/Im, is plotted in Figure 1
as a function of (spin speed)$^{-1/2}$. The spin cast films always have
higher Ie/Im values than the solvent cast films. Ie/Im increases
with spin speed, with larger increases observed for the higher molecular weight samples. Because the thickness of spin cast film is
proportional to (spin speed)$^{-1/2}$, Figure 1 can be also interpreted
as Ie/Im increases as film thickness decreases. The highest Ie/Im
was obtained for the 600,000 MW polystyrene sample spin cast at
8,000 rpm; this amounts to a 200% increase in Ie/Im compared with
the solvent cast film. The relaxation behavior of that film in the
region of the PS glass transition is plotted in Figure 2. As the
annealing temperature increased, Ie/Im dropped more rapidly and
attained lower final values after long annealing times. It is interesting to note that even after annealing for 24 hours the values
were still larger than those for solvent cast films.
    For a given annealing temperature the extent to which the spin
cast film has relaxed can be measured by the difference between
Ie/Im for a given annealing time and after 24 hours annealing. This
difference, calculated from Figure 2, is plotted as a function of
annealing time in Figure 3. A semilogarithmic relationship between
this difference in Ie/Im and annealing time was observed. Thus, it
is possible to calculate a relaxation rate constant for annealing
from the best fitted straight line. We found that films annealed at
higher annealing temperatures had a larger relaxation rate constant.
An Arrhenius plot of the relaxation rate constant vs the reciprocal
of the annealing temperature is shown in Figure 4. An activation
energy of 16.9 kcal/mol for the relaxation process was obtained.

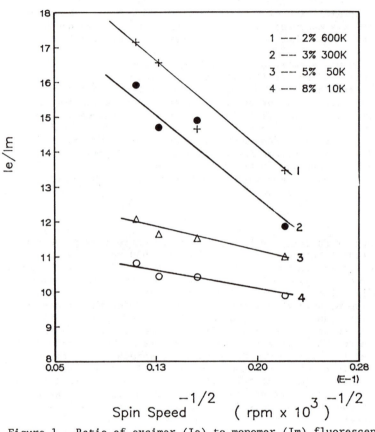

Figure 1. Ratio of excimer (Ie) to monomer (Im) fluorescence intensities of spin cast polystyrene films as a function of (spin speed)$^{-0.5}$.

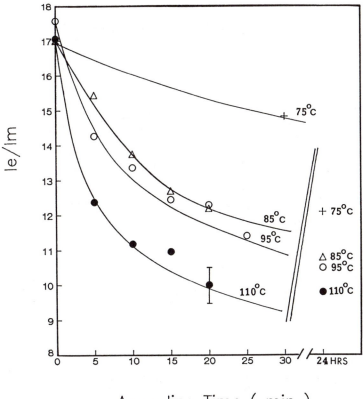

Figure 2. Time dependence for 600K MW polystyrene films spin cast at 8,000 rpm and annealed at different temperatures.

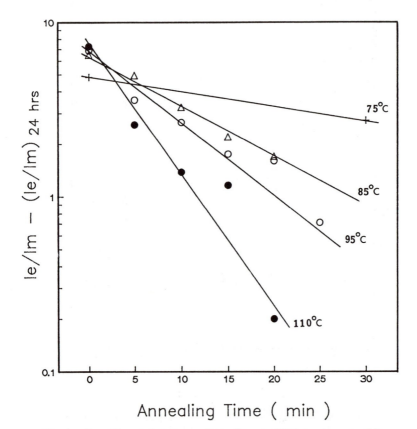

Figure 3.   The reduction of Ie/Im at different annealing
temperatures as a function of annealing time.

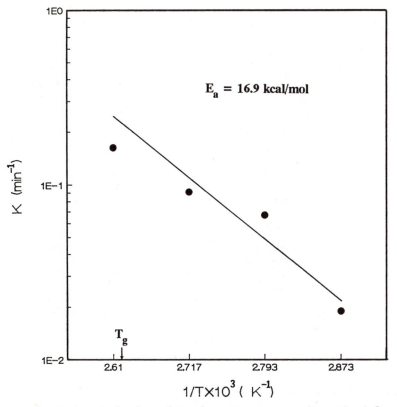

Figure 4. Arrhenius plot of rate constants determined for 600K MW polystyrene samples.

<u>Pyrene Fluorescence Spectra.</u> Solutions of the tagged polymers in
diglyme were prepared to have 1×10^{-6}M pyrene, based on the tagging
levels mentioned in the previous section. This resulted in polymer
concentrations of 2.33×10^{-4}, 8.47×10^{-5}, 3.65×10^{-5}, $1.99 \times$
10^{-5}, and 5.10×10^{-6}M monomer for the 0.43, 1.2, 2.7, 5.0, and
19.6% tagged polymers respectively. Emission spectra of the tagged
polymers were measured and compared to the spectra of pure PBA. The
emission spectra are plotted in Figure 5, normalized to the intensity of the emission at 398 nm. Ie/Im of the tagged polymers
increases with increasing tagging level and it is insensitive to
dilution (inset, Fig. 5), indicating that excimer formation in
solution is intramolecular.

    Solutions of the tagged novolacs in diglyme were spin cast into
films, and the spectra of these films were measured. Ie/Im of the
films was reduced significantly over that in solution (inset, Fig.
5), which is expected due to the extremely limited mobility of the
chromophores in the solid phase. Films were also prepared in which
the tagged polymer was "diluted" with untagged material. This
allowed the same overall concentration of pyrene to be achieved with
a different distribution in the film. The 19.6% tagged material was
combined with pure novolac to obtain films with concentrations
between 0.47 and 19.6 mol% pyrene. Films with various concentrations of free pyrene were also prepared. This allows the comparison
of three systems: 1) the pyrene is forced to be randomly distrib-
uted in the systems for which all novolac chains are tagged with
pyrene; 2) the pyrene is forced to be localized onto specific chains
in the blends of the 19.6 mole% pyrene tagged novolac mixed with
untagged novolac; and 3) the pyrene is free of direct constraints by
the polymer in novolac films containing free pyrene. Ie/Im for the
three systems are plotted in Figure 6. At pyrene concentrations
less than 6 mol%, Ie/Im is equivalent for the samples with free or
uniformly tagged pyrene (inset, Fig. 6). The films that are mixtures of tagged and untagged novolac have a much higher ratio of
excimer fluorescence at these low pyrene concentrations. At higher
pyrene concentrations, however, Ie/Im of the films with free pyrene
far exceeds that of the tagged pyrene.

DISCUSSION

<u>Spin Cast Polystyrene Films.</u> Ie/Im has previously been used to gain
information about chain conformation. Gupta and Gupta (3) studied
the fluorescence of uniaxially stretched polystyrene films, and
reported that 300% stretched polystyrene films have a 200% higher
ratio of excimer to monomer emission intensity. They suggested that
this was due to orientation effects which either increased the rotational mobility of the pendant groups, or promoted the formation of
excimer traps. However, in the glassy state the probability of
rotational mobility of the pendant groups is very low so that their
first explanation of the increase in Ie/Im is unlikely. It is known
that the orientation process in uniaxially stretched polystyrene
films produces alignment of the chains as well as an increase in the
concentration of trans conformational segments (20). This would be
expected to increase the number of intramolecular excimer forming
sites (EFS) formed between adjacent repeat units. However, that
increase amounted to only about 5% of the total trans conformational

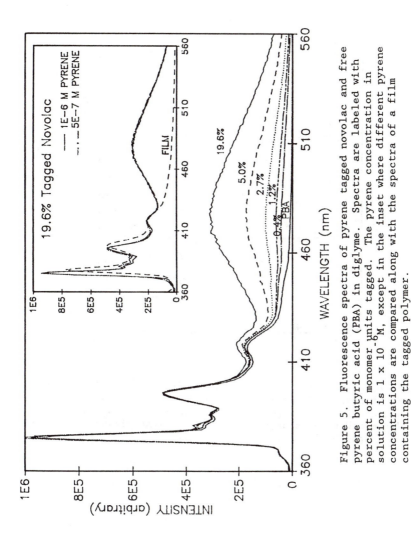

Figure 5. Fluorescence spectra of pyrene tagged novolac and free pyrene butyric acid (PBA) in diglyme. Spectra are labeled with percent of monomer units tagged. The pyrene concentration in solution is 1 x 10⁻⁶ M, except in the inset where different pyrene concentrations are compared along with the spectra of a film containing the tagged polymer.

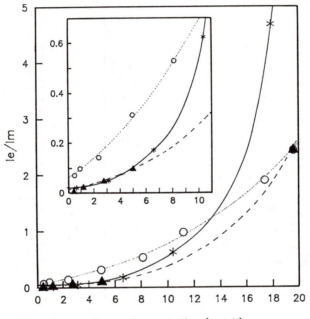

Figure 6.   Comparison of excimer intensity for free and tagged pyrene in novolac films.   Film composition: (*) free pyrene in novolac; (▲) pyrene tagged novolac; (o) 19.6 mol% pyrene tagged novolac mixed with untagged novolac.   The inset is an expansion of the graph for low pyrene concentrations.

segments for films stretched by 300% (3). This increase is insuffi-
cient to account for the 200% increase in Ie/Im observed for the
300% stretched films by Gupta and Gupta (3), as well as for the
difference between spin cast and solvent cast films observed in the
present study. It seems more likely that higher Ie/Im values of the
spin cast films can be explained by two factors: (1) the spin cast
films may have a higher concentration of intermolecular EFS; or (2)
there may be a higher rate of energy migration in the spin cast
films. As the film thickness decreases the polymer chain is more
likely to be aligned in the plane of the substrate which will
enhance those two factors.

The annealing experiments on spin cast films near Tg show a
drop in Ie/Im, although after 24 hours of annealing Ie/Im is still
higher than for a solvent cast film that has undergone no radial
stress. This implies that non-equilibrium chain structures still
exists in the spin cast films even after long annealing times.
Inter- or intra-molecular excimer formation between two chromophores
must result from rotation about methylene linkages; simple vibrational motion is not sufficient. Clearly, such rotation will be
hindered in the glassy state. However, short chain segmental motion
is possible below Tg even in the presence of chain entanglement, and
it could certainly lead to a change in the EFS population. Helfand
(21) has considered a number of fundamental cooperative segmental
motions that would produce the necessary trans/gauche rotations to
modify the number of EFS. In addition, Monnerie (22) has used various three bond and four bond jump models to simulate the dynamics
of motion on a tetrahedral lattice. Typically, the apparent activation energy for this secondary relaxation process is relatively
small, 10-20 kcal/mole (23), and only weakly dependent on side chain
structure. As we have shown in this work, the apparent activation
energy obtained from the kinetic measurements of the annealing process is also of this order. Thus, the experimental results are
consistent with this being a possible mechanism.

Pyrene / Novolac Films. Changes in Ie/Im can be used as a measure
of the local concentration of pyrene. In solution, excimers can be
formed by diffusion together of an excited and an unexcited molecule
during the lifetime of the excited state. At very low concentrations, little or no excimer formation is expected, as is the case
for the solution of pure PBA. If the pyrene is indeed tagged onto
the polymer chains, the local concentration of pyrene will be higher
in the vicinity of the chains for dilute solutions, and should result in higher Ie/Im. The increase in excimer emission with higher
levels of tagging, but equivalent overall pyrene concentration
(Figure 5), verifies that the pyrene is bound to the novolac.

In a film, however, molecular mobility is severely limited, so
that excimer fluorescence must arise mainly from pairs or groups of
pyrene molecules that were approximately in the excimer configuration when the film was cast. Thus, the intensity of the excimer
emission is also an indication of the local concentration of pyrene
in the cast film. If the pyrene aggregates, we expect that the
excimer fluorescence would increase with aggregation. This system
can be used to look at the aggregation of very low concentrations of
a small molecule dye in a polymer film, and potentially detect
molecular aggregation before it would be observable by other tech-

niques such as turbidity. In this case, the randomly tagged polymers are used as standards for molecularly dispersed pyrene. It was
found that the Ie/Im was nearly identical for free and tagged pyrene
at low dye concentrations (<6 mol%, or 10 wt%), but that the excimer
intensity increased rapidly for the free dye above that concentration, as shown in Figure 6, indicating aggregation of the free dye.
The novolac/pyrene system is similar to many positive photoresists
that have a small molecule dye (a diazoquinone) as a UV-sensitizer
dispersed in a novolac matrix. The dye/polymer compatibility may
reach a "saturated" level, as proposed by Law (16), above which
increased aggregation of the dye occurs. The concentration at which
this starts to occur for pyrene is well below the level of sensitizer loading for many of the positive resist systems currently in use
in industry. Such resists may contain as much as 15 to 50% sensitizer by weight (24,25).

Mixtures of the 19.6% tagged novolac with untagged novolac had
higher Ie/Im, at low overall pyrene concentrations, than the corresponding free or pure randomly tagged pyrene films - indicating that
intramolecular excimer formation is occurring for the 19.6% tagged
material. The differences in Ie/Im decrease with increasing pyrene
concentration, however, until the Ie/Im for the free pyrene sur-
passes that of the highly tagged material, indicating that the free
pyrene has aggregated to create local concentrations that are higher
than those enforced by the highly tagged material.

Summary

Previous research on spin cast films has centered on macroscopic
variables such as film thickness and uniformity. Our work has
focused on the microscopic properties of polymer chains that compose
these films, using materials and processing conditions that are
similar to those used in lithographic applications. We believe the
process of spin casting causes the polymer chains to exist in
oriented, non-equilibrium chain conformations. Orientation of the
polymer chains will increase the trans conformational segments as
found by Jasse and Koenig (20), which slightly increases the concentration of intramolecular EFS. However, this increase is insuffi-
cient to account for the observed 200% increase in Ie/Im caused by
the spinning process. An increase in intermolecular EFS and/or the
rate of energy migration must be included. After long annealing
times near the Tg the short chain stress can be relaxed by secondary
relaxation phenomena, as measured by Ie/Im value. Long chain
stress, however, still remains in the film. Consequently, significant residual stress may still remain in resist films after the
prebaking process commonly used in the industry. The spin cast
films of novolac polymers in solution with a small molecule dye
indicate that aggregation may occur for pyrene concentrations as low
as 6 mol%. This is well below the concentration of sensitizer in
many commercially used positive resist systems, indicating that the
resist films may be inhomogeneous.

Acknowledgments

This study was initiated with support from the Center for Materials
Research at Stanford University under the NSF-MRL program. Addi-

tional partial support was obtained from the Chemistry Division of the Office of Naval Research under contract N00014-87-K-0426. L. Kosbar would like to thank IBM for support through the Resident Study Program.

References

1. Elliott, D. J. Microlithography: Process Technology for IC Fabrication; McGraw-Hill: New York, NY, 1986; p 91.
2. Thompson, L. F.; Bowden, M. J. In Introduction to Microlithography; Thompson, L. F.; Willson, C. G.; Bowden, M. J., Eds.; ACS Symposium Series No. 219; American Chemical Society: Washington, DC, 1983; pp 161-214.
3. Gupta, M. C.; Gupta, A. Polym. Photochem. 1983, 3, 211-219.
4. Emslie, A. G.; Bonner, F. T.; Peck, L. G. J. Appl. Phys. 1958, 29, 858-862.
5. Acrivos, A.; Shah, M. G.; Petersen, E. E. J. Appl. Phys. 1960, 31, 963-968.
6. Meyerhofer, D. J. Appl. Phys. 1978, 49, 3993-3997.
7. Chen, B. T. Pol. Eng. Sci. 1983, 23, 399-403.
8. Weill, A. Springer Proc. Phys. 1986, 13, 51-8.
9. Jenekhe, S. A. Ind. Eng. Chem. Fundam. 1984, 23, 425-432.
10. Jenekhe, S. A. Polym. Mat. Sci. Eng. 1986, 55, 99-103.
11. Flack, W. W.; Soong, D. S.; Bell, A. T.; Hess, D. W. J. Appl. Phys. 1984, 56, 1199-1206.
12. Bornside, D. E.; Macosko, C. W.; Scriven, L. E. J. Imaging Tech. 1987, 13, 122-130.
13. Damon, G. F. In Proceedings of Second Kodak Seminar on Microminiaturization; Rochester, NY, 1967.
14. Daughton, W. J.; Givens, F. L. J. Electrochem. Soc. 1982, 129, 173-179.
15. Malangone, R.; Needham, C. D. J. Electrochem. Soc. 1982, 129, 2881-2882.
16. Law, K-Y. Polymer 1982, 23, 1627-1635.
17. Prest, W. M.; Luca, D. J. J. Appl. Phys. 1980, 51, 5170-5174.
18. Cohen, Y.; Reich, S. J. Polym. Sci., Polym. Phys. Ed. 1981, 19, 599-608.
19. Croll, S. G. J. Appl. Polym. Sci. 1979, 23, 847-858.
20. Jasse, B.; Koenig, J. L. J. Polym. Sci., Polym. Phys. Ed. 1979, 17, 799-810.
21. Helfand, E. J. Chem. Phys. 1971, 54, 4651-4661.
22. Valeur, B.; Jarry, J-P.; Geny, F.; Monnerie, L. J. Polym. Sci., Polym. Phys. Ed. 1975, 13, 667-674.
23. Frank, C. W. Macromolecules 1975, 8, 305-310.
24. Bowden, M. J. In Materials for Microlithography; Thompson, L. F.; Willson, C. G.; Frechet, J. M. J., Eds.; ACS Symposium Series No. 266; American Chemical Society: Washington, DC, 1984; pp 39-117.
25. Willson, C. G. In Introduction to Microlithography; Thompson, L. F.; Willson, C. G.; Bowden, M. J., Eds.; ACS Symposium Series No. 219; American Chemical Society: Washington, DC, 1983; pp 87-159.

RECEIVED July 13, 1988

Chapter 8

Effects of Various Additives on Accelerated Grafting and Curing Reactions Initiated by UV and Ionizing Radiation

Paul A. Dworjanyn, Barry Fields, and John L. Garnett

Department of Chemistry, University of New South Wales, Kensington, New South Wales 2033, Australia

The effect of mineral acid, specific inorganic salts such as lithium perchlorate and nitrate and organic compounds like urea as additives for accelerating a typical grafting reaction initiated by UV and ionising radiation are discussed. The model grafting system studied is styrene in methanol to polyethylene, polypropylene and cellulose. The role of polyfunctional monomers, particularly multifunctional acrylates in accelerating these grafting processes and also yielding synergistic effects with the preceeding additives is reported. The importance of monomer structure in these enhancement reactions is evaluated. A novel mechanism for these additive effects in grafting is proposed involving increased partitioning of monomer between grafting solution and substrate. The significance of this grafting work in the related fields of crosslinking and curing is discussed especially the role of the multifunctional acrylates in the presence of commercial additives such as silanes and fluorinated alkyl esters. The results indicate that by judicious choice of multifunctional acrylate in an oligomer mixture, concurrent grafting with cure can be achieved in both UV and EB rapid polymerisation systems.

The use of polyfunctional monomers (PFMs) such as trimethylol propane triacrylate (TMPTA) and divinylbenzene (DVB) as additives in accelerating grafting reactions initiated by UV and ionising radiation (1) has been reported. The implication of these results in analogous rapid curing processes has been discussed (2,3) including its relevance to the radiation crosslinking of monomers with PFMs (4). In the presence of mineral acids (4,5), specific inorganic salts (6) and organic compounds like urea (6), synergistic effects in both UV and ionizing radiation grafting have been observed with the above PFMs. Multifunctional acrylates (MFAs), of which TMPTA is a specific example, are special class of PFM which are frequently used in radiation rapid cure reactions.

0097–6156/89/0381–0112$06.00/0
© 1989 American Chemical Society

In the present paper, the effect of varying the structure of the MFA in the above grafting reactions is evaluated. In particular two monomer parameters have been investigated, namely the functionality and the effect of acrylate versus methacrylate groups. The significance of these MFA grafting results on analogous curing processes is discussed. A plausible mechanism to explain the data in both grafting and curing processes is proposed.

Experimental

Styrene was provided by Monsanto Chemicals (Australia) Pty.Ltd., whilst ethylene glycol dimethacrylate (EGDMA), diethylene glycol dimethacrylate (DEGDMA) triethylene glycol dimethacrylate (TEGDMA), polyethylene glycol dimethacrylate (PEGDMA), dipropylene glycol diacrylate (DPGDA), tripropylene glycol diacrylate (TPGDA), 1,6-hexane diol diacrylate (HDDA), trimethylol propane triacrylate (TMPTA), trimethylol propane trimethacrylate (TMPTMA), propoxylated glyceryl triacrylate (PGTA), hydroxy ethyl methacrylate (HEMA), ethoxy ethyl methacrylate (EEMA), tetrahydrofurfuryl methacrylate (TFMA), allyl methacrylate (AMA), N-vinyl pyrrolidone (NVP) and divinylbenzene (DVB) were obtained from Polycure Pty. Ltd., Sydney. All monomers were purified by column chromatography on alumina. The cellulose used was Whatman 41 filter paper, the polyethylene, low density, was supplied by Union Carbide as film (0.12mm) and polypropylene was isotactic double oriented film (0.06mm) ex-Shell.

Grafting Methods.

The procedures used for grafting were modifications of those previously reported (7). In the ionising radiation experiments grafting was performed in stoppered pyrex tubes (15 x 2.5cm) containing styrene solvent solutions (20ml) at room temperature, all reactions being carried out in quadruplicate. For the actual irradiations, polymer films or strips of cellulose of the appropriate size were fully immersed in the monomer solutions and the tubes irradiated in a 1200Ci cobalt-60 source. After completion of grafting the films were removed from the solution and solvent washed in a soxhlet extractor for 72 hours. When acid was used, especially for the cellulose runs, the films and paper strips were washed with methanol-dioxane (1:1) prior to soxhlet treatment. After removal of homopolymer, grafted polymers were dried to constant weight, the percentage graft being calculated as the percentage increase in weight of the grafted strip. With cellulose all strips were humidified to 65%r.h. at 20°C, prior to weighing, before and after grafting treatment.

In the UV studies, monomer solutions (20ml) were prepared in stoppered pyrex tubes in a manner similar to that described previously for the gamma irradiation work. Although evacuated quartz tubes give higher grafting efficiencies, reasonable rates of reaction were still achieved in the simpler pyrex system. For the irradiations, the tubes were positioned on a motor driven ventilated circulating drum at distances of 12-30cm from the UV source which was a 90W/Hg high pressure lamp. Experiments were performed at 24°C for the appropriate time as shown in the relevant tables. The polymer films were so positioned that during irradiation the surfaces of the films were perpendicular to the incident radiation. After irradiation, films and paper strips were treated as for the gamma system. Homopolymerisation was determined by the following procedure. At the completion of the irradiation, the grafting solution was poured into methanol (200ml) to precipitate the homopolymer and the sample tube rinsed with methanol (50ml). Homopolymer that adhered to both polymer film and tube was dissolved in dioxan (20ml) and the dioxan solution added to the methanol, in a beaker, together with benzene washings from the extraction of the original film. The beaker was heated to coagulate all polymer, the mixture cooled, filtered through a sintered glass crucible, washed with methanol (3 x 100ml) and the crucible dried to constant weight at 60°C. The percentage homopolymer was calculated from the weight of homopolymer divided by the weight of monomer in solution. The percentage grafting efficiency was then estimated from the ratio (graft/graft + homo-polymer) X100.

For the radiation rapid cure experiments, appropriate resin mixtures containing oligomers, monomers, flow additives and sensitisers (UV) were applied to the substrate as a thin coating, the material placed on a conveyor belt and then exposed to the UV and EB sources. The time taken to observe cure for each of the samples was then measured on a relative basis. The UV system used was a Primarc Minicure unit with lamps of 200W per inch. Two EB facilities were utilised namely a 500KeV Nissin machine and a 175KeV ESI unit.

Results.

Acid and PFM Additive Effects in Grafting Initiated by Ionising Radiation.

The general principles associated with radiation grafting including concepts such as Trommsdorff effects have previously been reviewed (7). In these earlier studies a number of additives have been reported to enhance grafting reactions initiated by ionising rad-iation (1,5,7). Typical of these additives are mineral acids (7) and polyfunctional monomers (PFMs) of which multifunctional acry-lates (MFAs) are the most frequently used (1). The data in Table I indicate typical enhancement results with these two additives for the radiation grafting of a representative model system, namely styrene in methanol to polyethylene film in the presence of gamma rays. The increase in grafting yields with both additives is particularly significant at the Trommsdorff peak, the shape of the enhancement curves being different for both sulfuric acid and TMPTA, the typical MFA used in the studies. When the two additives are combined in the same monomer solution, synergistic effects in radiation grafting are observed. This synergistic enhancement in yield is found at certain monomer concentrations, being particularly evident at the monomer concentration corresponding to the observat-ion of the gel peak.

Table I.   Synergistic Effect of Acid and TMPTA as Additives in
Grafting Styrene to Polyethylene Film Initiated by
Ionising Radiation[a]

Styrene	Graft (%)			
(% v/v)	N.A.	H^+	TMPTA	H^++TMPTA
20	14	19	-	-
30	37	51	39	54
40	76	81	73	106
50	109	134	137	181
60	89	73	105	101
70	68	62	59	95

[a] Dose rate of 4.1 x 10^4 rad/hr to 2.4 x 10^5 rad with
sulfuric acid (0.2M) and TMPTA (1% v/v); low density
polyethylene film (0.12mm); methanol solvent; N.A. =
no additive.

When TMPTA is replaced by a nonacrylate PFM such as DVB, analogous
synergistic effects with acid are observed as in Figure I where
polypropylene film is the substrate used. The data in this figure
also demonstrate the effect of dose rate at constant radiation dose
on the synergistic effect. Thus increasing the dose rate from
4.1 x 10^4 rad/hr to 1.12 x 10^5 rad/hr not only lowers the absolute
value of the grafting yield obtained in the presence of the two
additives from 210% to 43%, it also reduces the magnitude of the
graft enhancement. In addition to increasing grafting yields with
these two additives, there is also a synergistic effect observed
for the enhancement in grafting efficiency for the same grafting
reaction (Table II).

Hence the presence of acid and DVB not only increase grafting
yields, they also enhance competing homopolymerisation, however the
former reaction is preferentially affected to the benefit of the
overall process.

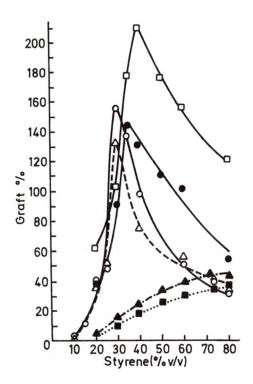

Figure 1. Role of dose rate in the synergistic effect of divinyl-benzene and sulfuric acid on grafting of styrene on polypropylene film in methanol   dose rate of 4.1 x 10^4rad/hr to total dose of 2.4 x 10^5rad: (△) styrene-methanol, (○) styrene-sulfuric acid (0.2M). (●) styrene-methanol-divinylbenzene (1% v/v), and (□) styrene-methanol-divinylbenzene (1% v/v)-sulfuric acid (0.2M).   dose rate of 1.12 x 10^5rad/hr to total dose of 2.5 x 10^5rad: (■) styrene-methanol, (▲) styrene-methanol-divinylbenzene (1% v/v)-sulfuric acid (0.2M).

Table II. Synergistic Effect of Acid and DVB on the Grafting
Efficiency of Styrene to Polyethylene Film Initiated
by Ionising Radiation[a]

Styrene (% v/v)	Grafting Efficiency (%)[b]		
	N.A.	H[+]	H[+]+DVB
30	56.2	59.4	58.8
40	73.9	83.0	74.2
50	75.1	79.4	85.2
70	45.7	41.2	65.4

[a] Dose rate of 1.0×10^4 rad/hr; N.A. = no additive; $H^+ = H_2SO_4(0.2M)$; DVB (1% v/v); other conditions as in footnote a, Table I.

[b] Ratio (graft/graft + homopolymer) X 100.

Comparison of Acid with Inorganic Salts and Urea as Radiation Grafting Additives.

Recent preliminary studies ($\underline{8,9}$) indicate that when specific inorganic salts such as lithium perchlorate and nitrate are used to replace mineral acid, analogous grafting enhancement is observed in the styrene polyethylene system, representative results being shown in Table III. Since this salt work, the first

Table III. Comparison of Acid with Inorganic Salts and Urea as
Additives in Grafting of Styrene to the Polyolefins
Initiated by Ionising Radiation[a]

Styrene Methanol (% v/v)	Graft (%)					
	Polyethylene[b]			Polypropylene[c]		
	N.A.[d]	H[+]	L	N.A.[d]	U	L
15	31	32	44	-	-	-
20	-	-	-	14	15	30
25	103	148	192	-	-	-
30	187	240	196	64	69	52
35	193	212	140	-	-	-
40	150	157	114	72	60	35

[a] $H_2SO_4(0.2M)$; $L=LiClO_4(0.2M)$; U= urea (0.2M)

[b] Irradiated at 3.3×10^4 rad/hr to dose of 2.0×10^5 rad.

[c] Irradiated at 7.5×10^4 rad/hr to dose of 2.5×10^5 rad.

[d] N.A.= no additive.

organic additives for enhancing radiation grafting yields have been discovered (6). Typical of these new additives is urea (Table III) which is more efficient than the lithium salt at certain monomer concentrations for grafting to polypropylene.

Consistent with the preceeding acid work, when TMPTA is added to the monomer solutions containing either lithium salt or urea, synergistic effects are observed for the radiation grafting of styrene to polypropylene (Table IV). Again the maximum increase in yield occurs in the monomer solution corresponding to the Trommsdorff peak.

Table IV. Synergistic Effect of TMPTA with Urea and Inorganic Salt Additives in Grafting Styrene to Polypropylene Initiated by Ionising Radiation[a]

Styrene methanol (% v/v)	N.A.	Graft (%) U+ TMPTA	L+ TMPTA
20	14	30(15)	29 (30)
30	64	135(69)	123 (52)
40	72	115(60)	83 (35)
50	44	75(41)	49 (23)
60	32	71(30)	15 (-)
70	26	48(24)	- (-)

[a] N.A.= no additive; U= urea (0.2M); L= $LiNO_3$(0.2M); TMPTA (1% v/v) ; irradiated at 7.5 x 10^4rad/hr to 2.5 x 10^5rad ; data in brackets without TMPTA

Additive Effects with Multifunctional Acrylates in UV Grafting.

When the source of initiation is altered from ionising radiation to UV, analogous additive effects to those previously discussed have been found. For reasonable rates of reaction, sensitisers such as benzoin ethyl ether (B) are required in these UV processes. Thus inclusion of mineral acid or lithium perchlorate in the monomer solution leads to enhancement in the photografting of styrene in methanol to polyethylene or cellulose (Table V). Lithium nitrate is almost as effective as lithium perchlorate as salt additive in these reactions (Table VI), hence the salt additive effect is independent of the anion in this instance. When TMPTA is included with mineral acid in the monomer solution, synergistic effects with the photografting of styrene in methanol to polyethylene are observed (Table VII) consistent with the analogous ionising radiation system.

Table V. Comparison of Acid with Inorganic Salts as Additives in Grafting Styrene to Polyethylene and Cellulose Initiated by UV[a]

Styrene methanol (% v/v)	Graft (%)					
	Polyethylene[b]			Cellulose[c]		
	N.A.[d]	H+	L	N.A.	H+	L
10	-	-	-	5	8	11
15	6	6	8	-	-	-
20	-	-	-	11	23	38
25	10	20	12	-	-	-
30	17	29	18	25	20	41
35	20	22	39	-	-	-
40	19	12	32	32	18	38

[a] H_2SO_4 (0.2M); L= $LiClO_4$ (0.2M)

[b] Irradiated 10 hr at 24cm from 90W lamp at 20°C, benzoin ethyl ether (1% w/v) sensitizer.

[c] Irradiated 15 hr as in footnote b.

[d] N.A. = no additive

Table VI. Effect of Lithium Nitrate as Inorganic Salt Additive in Grafting Styrene to Polypropylene Initiated by UV[a].

Styrene methanol (% v/v)	Graft (%)		
	B	L	B + L
20	5	1.9	10
30	24	1.4	38
40	45	1.7	20
50	31	2.0	14
60	22	4.5	13
70	15	4.1	11

[a] B = benzoin ethyl ether (1% w/v); L = $LiNO_3$ (0.2M); irradiated 8 hr with similar conditions to footnote b, Table V except 24°C.

Similar synergistic effects are found with TMPTA and the inorganic additive, lithium nitrate, for photografting to polypropylene (Table VIII).

Effect of Monomer Structure on Synergistic Effects with UV Grafting.

In terms of effectiveness of grafting enhancement with the MFAs TMPTA > HDDA > TPGDA ≈ DPGDA > PGTA. With the multifunctional methacrylates (MFMAs), a similar trend in photografting data is observed, TMPTMA being significantly better than DEGDMA, TEGDMA, EGDMA and PEGDMA (Table IX).

Table VII. Synergistic Effect of Acid and TMPTA as Additives in Grafting Styrene to Polyethylene Film Initiated by UV[a]

| Styrene | Graft (%) | | | |
(% v/v)	N.A.	H^+	TMPTA	H^++ TMPTA
20	28	14	28	41
30	101	126	52	78
40	189	193	321	266
50	124	107	412	525
60	37	31	133	188
70	32	39	-	-

[a] Irradiated 24 hr with similar conditions to footnote b, Table V. TMPTA (1% v/v); methanol solvent; N.A. = no additive.

Table VIII. Synergistic Effect of Multifunctional Acrylates with Inorganic Salts (L) in Grafting Styrene to Polypropylene Initiated by UV[a]

| Styrene in Methanol | Graft (%) | | | | | | | | | |
(% v.v)	T	T+L	P	P+L	D	D+L	TP	TP+L	H	H+L
20	100	126	50	55	18	47	17	94	53	115
30	297	529	113	147	112	162	99	217	81	239
40	496	337	289	146	370	235	336	187	370	180
50	283	221	131	79	191	81	167	98	315	91
60	281	81	98	46	114	70	118	57	113	67

[a] T = TMPTA, L = $LiNO_3$, P = PGTA, D = DPGDA, TP = TPGDA, H = HDDA.; all monomers (1% v/v); irradiation conditions as in footnote a, Table VI with B.

When the corresponding monofunctional monomers are used in these reactions, the overall level of photografting is significantly lower than with the MFMAs (Table X). The best of the monomers in Table X is AMA which is technically difunctional although a monomethacrylate. Interestingly, NVP also exhibits synergistic effects with lithium nitrate in the photografting of styrene in methanol to polypropylene, (Table X), although the level of reactivity is relatively low.

As the time of exposure to UV is increased, the magnitude of the synergistic effect between TMPTA and lithium nitrate is increased dramatically for the UV grafting of styrene to polypropylene (Table XI) such that after 16 hours of irradiation very large yields are obtained. Even with TFMA, the grafting yield in the 30% monomer solution is increased by almost one order of magnitude due to the synergistic effect.

Table IX.   Synergistic Effect of Multifunctional Methacrylates with Inorganic Salts(L) in Grafting Styrene to Polypropylene Initiated by UV[a]

Styrene in Methanol (% v.v)	Graft %									
	TM	TM+L	EM	EM+L	DM	DM+L	TEM	TEM+L	PM	PM+L
20	95	111	30	95	25	80	25	105	30	77
30	282	376	66	225	56	177	61	296	70	178
40	513	274	214	192	275	125	217	141	169	104
50	285	144	211	131	156	99	132	101	185	94
60	180	93	157	76	127	81	104	64	94	53

[a] TM =TMPTMA, EM = EGDMA, DM = DEGDMA, TEM = TEGDMA. PM = PEGDMA.; conditions as in footnote a, Table VIII.

Table X.   Synergistic Effect of Monofunctional Monomers Including Methacrylates with Inorganic Salts(L) in Grafting Styrene to Polypropylene Initiated by UV[a]

Styrene in Methanol (% v/v)	Graft (%)					
	AM	AM+L	EM	EM+L	N	N+L
20	36	59	5	10	8	10
30	136	305	15	63	25	45
40	331	88	60	39	60	20
50	98	40	53	27	20	12
60	-	-	40	13	-	-

[a] AM = AMA, EM = EEMA, N = NVP ; conditions as in footnote a, Table IX.

Synergistic Effects of TMPTA with Organic Additives (Urea, Silanes, Alkylester) on Photografting.

In Tables III and IV, the effect of urea on grafting initiated by ionising radiation was reported. Anologous results are obtained with the corresponding UV system (Table XII). The other important observation in Table XII involves the use of two commercial additives in the grafting process. These materials are used in

Table XI. Role of UV Dose in Synergistic Effect of Acrylate and Methacrylate Monomers with Inorganic Salts(L) in Photografting Styrene to Polypropylene[a]

Styrene in	Graft (%)									
Methanol	1 hr.			8 hr.			16 hr.			
(% v/v)	B	T	T+L	B	T	T+L	B	T	T+L	TF	TF+L
20	0.4	1.9	2.7	5	100	126	10	200	400	11	80
30	1.3	2.2	4.7	24	297	529	36	990	1820	38	280
40	1.9	3.3	7.4	45	496	337	116	1220	2100	181	118
50	1.7	7.2	7.4	31	283	221	53	460	530	112	65
60	-	-	-	22	281	81	40	810	660	76	27

[a] B = benzoin ethyl ether (1% w/v), T = TMPTA (1% v/v), L = LiNO$_3$(1% w/v), TF = TFMA (1% v/v); Graft at various exposure times of 1 hr, 8 hr and 16 hr.; other conditions as in footnote a, Table X

Table XII. Effect of Organic Additives (Urea, Silanes, Fluorinated Alkylesters) on Grafting of Styrene to Polypropylene Initiated by UV[a]

Styrene in	Graft (%)				
methanol (% v/v)	N.S.	B	B+U	B+U+Si	B+U+FE
20	<5	<5	<5	<5	<5
30	<5	35	30	18	23
40	<5	39	46	31	53
50	<5	17	19	13	16
60	<5	14	13	9	19
70	<5	14	11	7	10

[a] Irradiated 8 hr at 24cm from 90W lamp at 20°C; N.S. = no sensitizer; B = benzoin ethyl ether (1% w/v); Si = silane (1% v/v), Z-6020 supplied by Dow ; FE = fluorinated alkyl ester (1% v/v), FC-430 supplied by 3M.

industrial UV and EB rapid curing formulations to impart properties like slip and flow to the polymer film. Typical of these
industrial additives are the silane (Z-6020, ex-Dow) and the fluorinated alkyl ester (FC-430),ex.-3M). Inclusion of this silane in
the monomer solution for the photografting of styrene in methanol
to polypropylene leads to a retarding effect on the yield at all
monomer concentrations studied whilst the fluorinated alkyl ester
is an activator and leads to mild enhancement in UV grafting in the
same system. Most importantly, however, when TMPTA is added to the
monomer solution which already contains urea, silane and fluorinated
alkyl ester, a dramatic increase in grafting yield by almost two
orders of magnitude is observed (Table XIII). This synergistic

Table XIII. Effect of TMPTA in Presence of Organic Additives on
Grafting of Styrene to Polypropylene Initiated by UV[a]

Styrene in		Graft (%)		
methanol (% v/v)	N.S.	B	B+ U	B+ Additives + TMPTA
20	<5	<5	<5	260
30	<5	35	30	588
40	<5	39	46	711
50	<5	17	19	368
60	<5	14	13	283
70	<5	14	11	131

[A] Conditions as in Table XII ; additives used were urea,
silane and fluorinated alkyl ester ; TMPTA (1% v/v):
N.S. = no sensitizer.

TMPTA effect is unique since large enhancement in graft is achieved
even in the presence of retarders such as the silane. Finally it
is of interest to compare (3) the reactivities of the MFAs and
MFMAs in the present grafting work (Tables IX and XIII) with data
for the cross-linking efficiency of these same monomers with unsaturated olefins initiated by ionising radiation (Figure 2). The
results show that TMPTA is the best of the three MFAs examined in
this work, however TMPTMA is the best of the whole series for crosslinking.

TMPTMA > TMPTA > PGTA > TPGDA

Figure 2. Crosslinking efficiency of unsaturated olefins in the
presence of MFAs and MFMAs initiated by ionising radiation.

Discussion

The results of the present studies show that mineral acids, certain inorganic salts and organic compounds like urea are extremely useful additives for enhancing grafting yields in both photochemical and ionising radiation systems. The grafting profiles for each of these classes of additive are similar thus implying that common mechanistic pathways operate in each system. By contrast, the pattern of grafting with the PFMs, in particular the MFAs and MFMAs, is different from those of the preceeding three classes and thus the mechanism of the enhancement in grafting for the PFMs appears to be different. The fact that synergistic effects are observed when PFMs are added to the monomer solutions containing the various additives is consistent with the conclusion that different mechanistic pathways in grafting enhancement operate for the PFMs when compared with the other additives.

Mechanism of the Acid Effect in Grafting. Clarification of the mechanism of the acid effect in grafting initiated by UV or ionising radiation has been complicated by the variety of chemical constituents present in any one grafting system ($\underline{7}$, $\underline{10\text{-}14}$). Thus the inclusion of solvents, monomers, backbone polymers and any other materials in the grafting solution such as sensitisers (for UV) need to be considered in any overall treatment. Solvents are particularly relevant since solvents which wet and swell the backbone polymer generally enhance grafting ($\underline{15}$). Originally the acid effect was observed in grafting initiated by ionising radiation ($\underline{7}$). At that time the enhancement due to acid was attributed essentially to two predominant factors, namely the radiolytic yield of hydrogen atoms and also the extent to which grafting polymer (polystyrene) was solubilised in the bulk solution. Solvent structure was also considered to be important, those solvents with high radiolytic yields of hydrogen atoms, such as methanol and the lower molecular weight alcohols, being efficient reagents for grafting. This observation was explained in terms of the abstraction of hydrogen atoms from the trunk polymer PH by those H atoms yielding additional grafting sites as depicted in Equations (1) to (3).

$$CH_3OH + H^+ \rightarrow CH_3OH_2^+ \qquad (1)$$
$$CH_3OH_2^+ + e \rightarrow CH_3OH + H \qquad (2)$$
$$PH + H \rightarrow P^\cdot + H_2 \qquad (3)$$

The above mechanism has proved to be an oversimplification of the grafting process since it fails to explain a number of further observations recently reported ($\underline{7}$, $\underline{8}$, $\underline{16}$, $\underline{17}$) including (i) the occurrence of enhancement only at certain acid and monomer concentrations, (ii) the efficacy of H_2SO_4 and HNO_3 over other acids, (iii) the presence of grafting enhancement in the preirradiation technique where radiolytically produced hydrogen atoms are not available and (iv) acid enhancement in sensitised photografting (Table V) where radiation chemistry effects are not relevant.

Mechanistically one of the most important observations concerning the acid effect is the recent detailed systematic study of the swelling of polyethylene in the presence of methanolic solutions of styrene (18) which indicate that partitioning of non-polar monomer into non-polar substrates may be significantly improved (Table XIV) by the inclusion of mineral acid in the grafting solution. Styrene labelled with tritium was used for these sophisticated experiments which indicate that most swelling occurs within the first minute of exposure of backbone polymer to solution. Based on these data and more extensive evidence recently reported (8,9), a new model has been proposed to explain the effect of acids in enhancing radiation grafting. In any grafting system at any one time, there is an equilibrium concentration of monomer absorbed within the grafting region of the backbone polymer. This grafting region may be continually changing as grafting proceeds. Thus in grafting styrene to cellulose during the initial part of the reaction, the grafting region will be essentially cellulosic in nature, however, as reaction proceeds, the grafting region will become more styrenated. The degree to which monomer will be absorbed by this grafting region will therefore depend on the chemical structure of the region at the specific time of grafting. Experimental data similar to that shown in Table XIV indicate that increased partitioning of monomer occurs in the graft region when acids are dissolved in the bulk grafting solution. Thus higher concentrations of monomer are available for grafting at a particular backbone polymer site in the presence of these additives. The extent of this improved monomer partitioning depends on the polarities of monomer, substrate and solvent and also on the concentration of acid. It is thus the effect of these ionic species on partitioning which is essentially responsible for the observed increase in radiation grafting yields in the presence of acid additives. Radiolytically generated free radicals would be expected also to make some contribution to the acid effect in a system where initiation

Table XIV. Variation in Styrene Absorption from Methanol Solution by Polyethylene with Time[a]

Time of Swelling (h)	Styrene Absorption (mg styrene / g polyethylene)	
	N.A.	H_2SO_4(0.1M)
0	0	0
0.5	38.8	42.8
1	40.5	44.3
2	42.5	45.8
12	45.6	53.3
20	45.4	54.0

[a] Technique used involved tritiated styrene (18) with 30% styrene in methanol solutions (% v/v) and polyethylene film (0.12mm) at 25°C; N.A. = no additive.

was by ionising radiation, however this would not be the predominant
pathway for the enhancement. The above model developed for grafting
by ionising radiation is also applicable to analogous sensitised
photografting and would explain the large increases in yield observed
in the presence of acid in the latter system (Table V).

Mechanism of Additive Effects with Inorganic Salts and Urea.

The data in Tables III, V and VI indicate that inorganic salts
such as lithium perchlorate and lithium nitrate can also enhance
grafting reactions initiated by UV or ionising radiation. When acid
and salt are compared as additives (Table III), in the grafting of
styrene in methanol to polyethylene using ionising radiation, the
salt enhancement only occurs at the lower monomer concentrations
whereas, with acid, increased grafting yields are observed in all
styrene concentrations studied. With the corresponding UV system
(Table V), the reverse is observed for grafting to both polyethylene
and cellulose. Thus in the presence of the salt, photografting
yields are increased at all monomer concentrations used whereas with
acid the enhancement only occurs in the lower concentration monomer
solutions. Mechanistically it is proposed that the salt effect is
similar to acid i.e. protons are not essential to the operation of
the graft enhancement and any small specific differences between the
two type additives can be attributed to differences in polarities
within each system. The behaviour observed can be reconciled with
an increased availability of monomer within the graft region in the
presence of metal salt.

Consistent with the above acid and salt additive effects is the
use of organic compounds, typified by urea, for enhancing both photografting and radiation grafting yields (Table III, XII and XIII).
A similar explanation is advanced to explain the urea effect namely
the occurrence of increased partitioning of monomer between solution
and substrate in the presence of additive.

Mechanism of Polyfunctional Monomer as Grafting Additives.

Inclusion of PFMs such as TMPTA (Tables I and VII) and DVB
(Figure 1) in additive amounts ($\approx$ 1%), significantly enhance grafting
in both UV and ionising radiation systems, especially at the Trommsdorff peak. The PFMs appear to have a dual function (Figure 3),
namely to enhance the copolymerisation and also crosslink the grafted
polystyrene chains. In the grafting experiments, branching of the
growing grafted polystyrene chains occurs when one end of the polyfunctional monomer (e.g. DVB) immobilised during grafting is bonded
to the growing chain. The other end is unsaturated and free to
initiate new chain growth via scavenging reactions.

The new branched polystyrene chain may eventually terminate,
crosslinked by reacting with another polystyrene chain or an immobilised divinylbenzene radical. Grafting is thus enhanced mainly

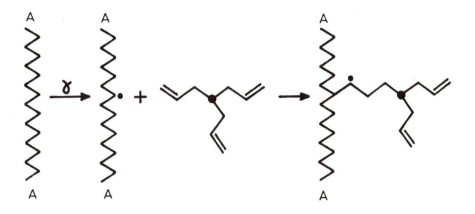

Figure 3. Interaction of multifunctional acrylate with polymer
radical by gamma irradiation.

through branching of the grafted polystyrene chain. A comparison
of the UV and ionising radiation results indicates that the magnitude of the increase in grafting yield is significantly higher in
the photochemical system particularly with the MFAs, thus suggesting
that in the UV the acrylate group is acting as an additional sensi-
tiser for the grafting process.

Synergistic Effects of PFMs with Acids, Salts and Urea.

Although the yields of copolymer in the presence of the PFMs
are similar in magnitude to those when the other additives such as
acid are used, the shapes of the grafting curves (Figure 1, DVB as
typical PFM) with the two types of additives are different. This
conclusion is consistent with the different mechanisms proposed for
each type additive and suggests that both additives may be used
simultaneously in solution to give a synergistic effect and increase
the radiation grafting yield further. The data in Figure 1 confirm
this conclusion and show that the inclusion of both acid and DVB
result in very large increases in graft. The presence of both
additives also enhances homopolymerisation, however grafting efficiency is improved under these conditions (Table II) and the grafting

process is preferentially favoured at certain monomer concentrations. The efficiency is particularly high at 50% styrene concentration.

In addition to yielding synergistic effects with acids in both photografting and radiation grafting processes, the PFMs exhibit analogous synergism with inorganic salts (Tables III, IV, VIII, IV, XI) such as lithium perchlorate and organic additives like urea (Tables IV, XII). Again with these last two classes of additive, the enhancement due to the synergistic effect reaches a maximum at the Trommsdorff peak which is the region where the length of the grafted chains is also a maximum. Thus, in addition to increasing the grafting yields, the presence of these additives may also influence the nature and structure of the copolymer formed.

Effect of Structure of PFM on Enhanced Grafting.

Of the range of MFAs and MFMAs studied in this work (Tables VIII and IX), TMPTA and TMPTMA demonstrate the highest enhancement in photografting, their grafting profiles being almost coincident. Although both acrylates are present in 1% v/v concentration, TMPTMA has a higher molecular weight, thus it is more effective on an equivalent basis than TMPTA in increasing yield. This difference may reflect the higher hydrocarbon functionality of the TMPTMA, leading to higher absorption of monomer into the backbone polymer (i.e. increased partitioning) resulting in enhanced reactivity.

Addition of lithium nitrate to solutions separately containing ITMPTA and TMPTMA leads to a splitting of the previously coincident profiles. This result may indicate an interaction between the lithium ion and the acrylates yielding charge-transfer complexes possessing differing sensitisation efficiencies in the UV. Analogout types of charge-transfer complexes are known between macrocyclic ethers and lithium ions. Such complex formation may also contribute to the synergistic effects currently observed in grafting when lithium salts with MFAs (and MFMAs) are both present in the same monomer solution.

The significance of the acrylate functionality in these enhanced grafting processes is further demonstrated by the data in Table X where the monofunctional EEMA is much less reactive than any of the MFAs or MFMAs in the preceeding two tables. Inclusion of an olefinic bond in the appropriate position in the methacrylate monomer increases the reactivity of the monomer as shown by AMA in Table X. As the time of irradiation is increased, the effect of functionality on the grafting yields becomes much more significant (Table XI) such that after 16 hours there is an order of magnitude difference between TMPTA and TFMA.

Preparative Significance of Current Additive Effects in Grafting.

In addition to being of fundamental importance to basic radiation grafting theory, the results with the range of additives previously discussed are of value in a preparative context since radiation graft copolymers are now being used for a wide range of applications and any technique for reducing the radiation dose to achieve a particular percentage graft is of both practical and economic significance. The discovery of a range of metal salts and organics, such as urea, to complement acid already known, as additives for enhancing grafting is valuable since the scope of the enhancement technique is very muc·h expanded. Thus grafting systems which may have been sensitive to acid, can now be treated with metal salts or urea to achieve the same type of enhanced reactivity. By contrast, if solubility problems in the grafting solution occur with the latter two additives, acid can be used as an alternative.

The observation that polyfunctional monomers, in particular multifunctional acrylates and methacrylates, exhibit dramatic synergistic effects with these additives is also of great preparative significance in grafting. Such a development will be particularly beneficial for modifying surfaces of relatively inert materials where generally relatively high radiation doses (5 megarads) are required to achieve even low copolymerisation yields (19). Under these relatively high radiation doses, not only can the structure of the backbone polymer be detrimentally affected, but also internal crosslinking of the grafted copolymer can occur to yield a surface which is unsuitable for many applications. By the use of the current PFM and MFA synergistic effect, the radiation dose required to achieve a particular percentage graft can be significantly reduced to levels where no adverse radiolytic effects in the finished copolymers are observed.

Relevance of Current Additive Effects in Curing Reactions.

The present data are important in radiation rapid cure (RRC) reactions where films of oligomer/monomer are polymerised in a fraction of a second using high pressure UV lamps and low energy electron beam (EB) machines. In these RRC applications, MFAs are used for two predominant purposes, namely to accelerate rate of polymerisation and achieve crosslinking of the cured film. RRC formulations also contain additives to control slip, gloss, flow etc., and these are typified by the fluorinated ester and the silane used in the present study (Table XII). When these commercial additives are included in a monomer grafting solution, the results show that the surface active fluorinated compound enhances graft whereas the silane is a retarder, presumably due to the repulsion effect of the silicon atom. However the most significant result with these additives was the TMPTA data where dramatic increases in graft were observed at low TMPTA concentrations despite the presence of the silane retarder (Table XIII). The mechanistic role of MFA in RRC data where dramatic increases in graft were observed at low TMPTA concentrations would thus appear to be more subtle than hitherto considered, TMPTA not only speeds up cure and crosslinking, it can also markedly affect the occurrence of concurrent grafting during cure. Thus judicious choice of MFA in RRC mixtures could lead to graft enhancement during cure with beneficial properties to the finished product.

The reactivity of TPGDA in the present grafting work is also relevant to RRC reactions. TPGDA is one of the most frequently used monomers in RRC formulations since its properties are a good compromise between speed of cure, viscosity reduction of oligomers and low Draize value. As an MFA additive in the current grafting work, it is less efficient than TMPTA. A similar trend in reactivity of TPGDA in RRC curing (Figure 4) and crosslinking processes (Figure 2) is also found (3, 21). Interestingly, in terms of cure speed, NVP is very much faster than any of the MFAs studied in this current work (Figure 4), however as an accelerator in grafting, NVP is relatively poor (Table X). Thus, even though NVP is fast to cure, this monomer possesses other deficiencies which indicate that, as a compromise in overall properties, MFAs such as TPGDA are the preferred reactive diluents in many RRC formulations.

$$NVP\ 1.00 > TMPTA\ 0.56 > PGTA\ 0.41 >$$
$$TPGDA\ 0.33 \approx HDDA\ 0.33 > DPGDA\ 0.26$$

Figure 4. Relative Rates of Curing of MFAs Compared with NVP in admixture (1/1, v/v) with Urethane Acrylate and ß (1% w/v) using UV Initiation (20).

Acknowledgments

The authors thank the Australian Institute of Nuclear Science and Engineering, the Australian Research Grants Committee, and the Australian Atomic Energy Commission for financial assistance.

Literature Cited

1.   Ang, C.H.; Garnett, J.L.; Levot, R.; Sangster, D.F. J.Appl. Polym.Sci. 1982, 27, 4893.

2.   Bett, S.J.; Garnett, J.L.; Proc.Radcure Europe '87 of SME Munich., 1987, in press.

3. Dworjanyn, P.A. ; Garnett, J.L. Rad.Phys.Chem., in press.

4. Micko, M.M.; Paszner, L. J.Rad.Curing. 1974, 1(4), 2.

5. Ang, C.H.; Garnett, J.L.; Levot, R.; Long, M.A. J.Polym.Sci. Polym.Lett. 1983, 21, 257.

6. Dworjanyn, P.A.; Garnett, J.L. J.Polym.Sci.Polym.Lett., in press.

7. Garnett, J.L. Rad.Phys.Chem. 1979, 14, 79.

8. Garnett, J.L.; Jankiewicz, S.V.; Long. M.A.; Sangster, D.F. J.Polym.Sci.Polym.Lett. 1985, 23, 563.

9. Garnett,J.L. ; Jankiewicz, S.V.; Levot,R. ; Sangster,D.F. Rad. Phys.Chem. 1985, 25, 509.

10. Gecintov, N.; Stannett, V.T. ; Abrahamson,E.W. ; Hermans, J.J. J.Appl.Polym.Sci. 1960, 3, 54.

11. Arthur, J.C. Jr. Polym.Preprints 1975, 16, 419.

12. Kubota, H.; Murata, Y.; Ogiwara, Y. J.Polym.Sci. 1973, 11, 485.

13. Tazuke, S.; Kimura, H. Polym.Lett. 1978, 16, 497.

14. Hebeish, A.; Guthrie, H.T. The Chemistry and Technology of Cellulocis Copolymers; Springer-Verlag: Berlin Heidelberg, 1981.

15. Odian, G.; Sobel, M.; Rossi, A.; Klein, R. J.Polym.Sci. 1961, 55, 663.

16. Chappas, W.J.; Silverman, J. Rad.Phys.Chem. 1979, 14, 847.

17. Hoffman, A.S.; Ratner, B.D. Rad.Phys.Chem. 1979, 14, 831.

18. Garnett,J.L.; Jankiewicz, S.V.; Sangster, D.F. J.Polymer Sci. Polym.Lett., in press.

19. Garnett, J.L. ; Phuoc, D.H. unpublished work.

20. Betts, S.J. ; Garnett, J.L. unpublished work.

21. Smith, B.M. ; Garnett, J.L.; Burford, R.P. unpublished work.

RECEIVED July 13, 1988

Chapter 9

Radiation Chemistry of Polymers for Electronic Applications

Elsa Reichmanis

AT&T Bell Laboratories, Murray Hill, NJ 07974

Polymer chemistry and polymer radiation chemistry in particular are key elements of the electronics industry. Polymer materials that undergo radiation induced changes in solubility are used to define the individual elements of integrated circuits. As the demands placed on these materials increases due to increased circuit density and complexity, new materials and chemistry will be required. Many of the new chemistries that are being developed are described in this article.

A key step in the manufacture of integrated circuit devices involves the definition of a circuit pattern into a radiation sensitive polymeric material called a resist (1). This aspect of the microlithographic process is outlined in Figure 1. The desired semiconductor substrate is first coated with a thin layer ($\sim$0.5-2.0 μm thick) of a resist that is then exposed to some form of radiation in an image-wise manner. The radiation sensitive moieties of the resist undergo reaction to generate the desired circuit pattern in the polymeric material. This pattern may then be developed into a relief image through treatment with an appropriate solvent. If exposure to radiation results in reduced solubility or crosslinking, the material is classified as a negative resist. Alternatively, a positive resist undergoes radiation-induced reactions that lead to enhanced solubility in the exposed regions. After definition of the circuit pattern into the resist layer, the features are transferred into the substrate by either a wet-, or dry-etching technique. Alternatively, dopants such as arsenic or boron could be diffused into the exposed regions to afford the desired electrical characteristics. The resist is then stripped from the substrate. Subsequent dielectric, insulator or conducting materials are then deposited onto the surface of the device substrate and patterned in a similar manner. A succession of lithographic processes leads to a completed integrated circuit device.

The radiation sources employed in microlithography include conventional (>300 nm) and deep-UV (<300 nm) light, electron-beam, ion-beam and x-ray sources. By far the predominant lithographic technology is conventional photolithography which

0097–6156/89/0381–0132$06.75/0

○ 1989 American Chemical Society

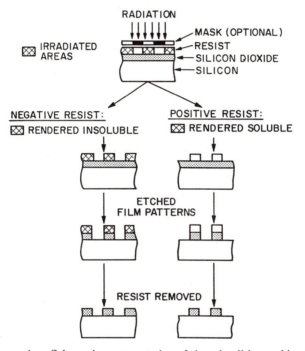

Figure 1   Schematic representation of the microlithographic process.

employs 350-450 nm radiation. Improvements in the printing technology have allowed a reduction in feature size of state of the art devices from 5 to 6 μm 1976 to less than 1 μm today. It is anticipated that conventional photolithography will be able to print 0.6-0.8 μm features and will continue to be the dominant technology in the near term. The use of deep-UV excimer laser sources is expected to push the limits of optical lithography into the sub-0.5 μm regime. Several step-and-repeat 5 or 10X reduction systems have been built that use 248 nm KrF excimer laser sources (2). These sources provide satisfactory intensities to accommodate resists with 50-150 mJ cm^{-2} sensitivities. The alternatives to optical lithography include x-ray, and scanning or projection electron and ion beam lithography (3). Regardless of which technology becomes dominant after photolithography has reached its limit, new resists and processes will be required.

While sensitivity to a given radiation source that is commensurate with the desired throughput is a key resist requirement, other materials properties must also be considered. Since resists are generally coated onto a substrate by spin-coating techniques, the polymeric material must be soluble in a suitable solvent. After irradiation, it must exhibit differential solubility between the exposed and unexposed regions to facilitate pattern definition. A resist must also exhibit good adhesion to materials commonly used in semiconductor device manufacturing. These include metals such as aluminum, chromium, and gold, in addition to materials such as silicon or silicon dioxide. If a given resist fails to properly adhere to the substrate, loss of resolution or more critically, loss of the entire circuit pattern will result. Etching resistance is another key materials characteristic that must be considered in the design of resist chemistry. Clearly, a given material must withstand the etching environment that is used to transfer the resist image into the device substrate.

The trends towards increased device complexity and decreased feature size necessitate that any given resist material for future lithographic processes be capable of sub-0.5 μm resolution. The resolution that a given resist can achieve is largely determined by resist contrast (γ), a parameter that can be directly related to the radiation chemistry the material undergoes. In the case of a negative resist, contrast is related to the rate of cross-linked network (gel) formation. For a positive resist, contrast relates to both the rate of polymer degradation and/or the rate of change in solubility of the material. The value of γ is obtained from the slope of the linear portion of the exposure response curves shown in Figure 2. Generally, high contrast is expected to minimize effects such as scattering of radiation in a resist film, thus allowing higher resolution imaging.

Sensitivity is another resist property that relates directly to the radiation chemistry of the system. Fundamentally, the intrinsic radiation sensitivity of a material will determine its lithographic response to radiation. For a photochemical reaction, this is defined as the quantum yield (Φ) of photoreaction, where Φ is the number of photo-induced events per number of photons absorbed. For high energy radiation processes, the parameter of interest is the G value which is defined as the number of radiation-induced events per 100 eV of absorbed energy. The chemistry associated with positive resist systems should exhibit high values of G_s, or $G_{scission}$, with little or no cross-linking. Alternatively, negative resists should possess high values of $G_{crosslinking}$ (G_x) with negligible scission. In both cases, optimization of other molecular parameters such as molecular weight, molecular weight distribution and glass transition temperature (T_g) is also required to effect optimum resist performance.

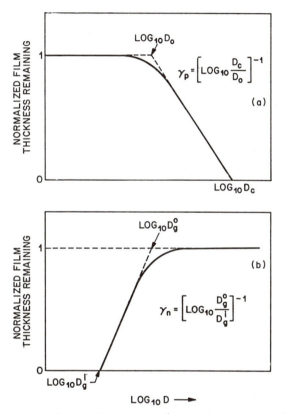

Figure 2   Typical lithographic response (contrast) curves for (a) positive and (b) negative resists.

Optimization of both resist sensitivity and contrast requires a fundamental appreciation of the radiation chemistry in addition to appreciation how polymer molecular parameters affect the lithographic behavior of the resist. The intent of this chapter is to further the readers understanding of the polymer and radiation chemistry that is associated with a large part of the microelectronics industry, and provide some of the necessary background to effect future developments.

1. CONVENTIONAL NEGATIVE RESIST CHEMISTRY

As defined above, negative resists are those materials which become insoluble after exposure to radiation. Generally, they undergo some form of radiation induced crosslinking whereby the crosslinking component is either an integral part of the polymer or is incorporated into the polymer as an additive. Development, or removal of the unirradiated portions of the film, occurs *via* swelling of the polymer matrix followed by chain disentanglement and dissolution. Since this often leads to swelling of the resist in the nominally crosslinked regions, the resolution capability of negative resists is often limited. However, Novembre and co-workers have shown that use of the Hansen 3-dimensional solubility parameter model which provides a thermodynamic solubility picture of a resist, aids in the determination of a developer that will only minimally contribute towards resist swelling during development (4). Thus, developer selection and high resolution imaging in negative resists is facilitated. Some general characteristics of negative resists include high radiation sensitivity, good adhesion and good dry-etching resistance.

Probably the most well known and widely used negative resists are formulations of cyclized rubbers with *bis*-arylazide crosslinking agents (Figure 3) (5,6). The matrix resin is typically a cyclized poly(cis-1,4-isoprene) that is soluble in a wide range of non-polar organic solvents, may be spin coated to generate uniform films, and exhibits excellent adhesion to a variety of substrates. Addition of the *bis*-arylazide effects crosslinking and insolubilization after UV irradiation. The mechanism of this reaction involves generation of a nitrene intermediate followed by reaction of this nitrene with the cyclized rubber matrix (Figure 3).

A major limitation of these resists is their dependence on organic solvent developers that cause image distortion due to swelling. Workers at Hitachi have developed an alternate aqueous developed, two component resist that attempts to overcome this problem (7). The resist, called MRS, utilizes an aqueous base soluble phenolic resin such a poly(p-vinylphenol) as the matrix resin, and a photosensitive crosslinking agent such as 3,3'-diazidodiphenyl sulfone. Irradiation results in the formation of a crosslinked network *via* a route similar to the cyclized rubber resists. Presumably, nitrene insertion occurs at the backbone carbon-hydrogen bonds since this mechanism is energetically favored over insertion into the aromatic ring. While the solubility of the resist in aqueous developers decreases with UV irradiation and no swelling is observed during development of sub-micron images, the development step must be tightly controlled to minimize feature undercut and maintain linewidth control. This is because the high optical density of the resist in the exposure range (250-320nm) causes most of the photochemistry to take place in the surface layers of the resist. The result is that a crosslinking gradient occurs through the thickness of the film and material near the resist-substrate interface is believed to be minimally affected by irradiation.

The realm of negative resist chemistry also includes many materials that incorporate the crosslinking unit into the polymer by covalent bonding. In these cases, the radiation sensitive unit is an integral part of the polymer appearing in either the backbone or side-chain of the material. These systems frequently crosslink by a chain mechanism that leads to high crosslinking efficiency and high resist sensitivity. The crosslinking unit is typically incorporated into the polymer through the side chain and examples of useful materials include vinyl, epoxy and halogen containing resins.

Examples of vinyl containing polymers that have been employed as negative resists include poly(diallyl) ortho-phthalate)(PDOP) (8) and poly(allylmethacrylate-co-2-hydroxyethyl meth-acrylate) (9). These materials, and others of similar structure (10,11), take advantage of the propensity of carbon-carbon double bonds to undergo radiation induced polymerization. Crosslinking occurs through the unsaturated side-chain and high electron-beam sensitivity has been reported. The epoxy moiety is another unit that is well known to undergo efficient crosslinking reactions. The chemistry involves a chain mechanism that leads to highly sensitive resists with high crosslinking efficiencies. However, the crosslinking reaction continues after exposure leading to post-exposure curing. This results in a growth in feature size that is dependent upon the time the material remains in vacuo after exposure. Since the reaction is diffusion controlled, the extent of reaction is also dependent on the T_g of the resist. Examples of epoxy containing negative resists include epoxidized 1,4-poly(butadiene) (12,13) and poly(glycidyl methacrylate-co-ethyl acrylate)(COP) (14).

The incorporation of halogen into acrylate and styrene based polymers has been found to instill high sensitivity to radiation induced crosslinking. The proposed mechanism involves radiation induced cleavage of the carbon-halogen bond to generate a radical that may then undergo rearrangement, abstraction or recombination reactions to afford a crosslinked network (Figure 4) (15). The localized nature of the reaction in these polymers, in addition to the generally higher values of T_g, eliminates curing effects that are found in the epoxy and vinyl containing resists. An example of a halogen containing, negative, X-ray resist is poly(2,3-dichloropropylacrylate)(DCPA) (16). The sensitivity of DCPA is largely due to the enhanced sensitivity of the carbon-halogen bond to radiation induced cleavage in addition to an increase in X-ray absorption. Chlorinated styrenes have proved useful in the development of negative e-beam and deep-UV resists. Poly(chlorostyrene-co-glycidyl methacrylate) (GMC) (17,18) exhibits a sensitivity in the range of 1-5 μC cm^{-2} at 20 kV, good resist contrast ($\sim$ 1.4) and sub-micron resolution (Figure 5). Alternatively, incorporation of chloromethyl groups into polystyrene is another means of improving sensitivity to radiation induced crosslinking (19,20).

Polymer molecular properties such as molecular weight and polydispersivity have a significant effect on the lithographic behavior of the single component negative resists described above. For example, it has been shown for a series of poly(chloromethylstyrene) (PCMS) polymers that an increase in polymer molecular weight by a factor of about 10 results in a ten-fold increase in resist sensitivity (21,22). Though resist contrast is unaffected, resist resolution decreases with increasing molecular weight. Resist contrast will however be affected by polymer dispersivity; decreased dispersivity is expected to lead to increased contrast and improved resolution.

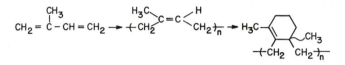

CYCLIZED POLYISOPRENE BASE RESIN

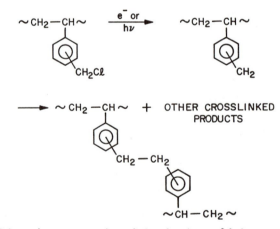

BIS-AZIDE PHOTOCHEMISTRY

Figure 3   Schematic representation of the chemistry of the cyclized rubber - *bis* arylazide negative resists.

~ CH$_2$—CH ~ $\xrightarrow[h\nu]{e^- \text{ or}}$ ~ CH$_2$—CH ~

⟶ ~ CH$_2$—CH ~   +   OTHER CROSSLINKED PRODUCTS

Figure 4   Schematic representation of the chemistry of halogenated styrene based negative resist.

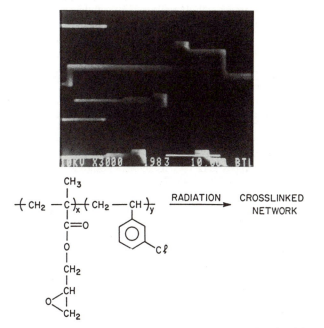

Figure 5   SEM micrograph depicting 0.75μm imaging in GMC with a schematic representation of GMC chemistry.

2. CONVENTIONAL POSITIVE RESIST CHEMISTRY

Materials that exhibit enhanced solubility after exposure to radiation are defined as positive resists. Positive acting materials are particularly attractive for the production of VLSI devices because of their high resolution properties. The chemistry of these systems generally involves either chain-scission or solution-inhibition mechanisms.

The workhorse of the VLSI industry today is a composite novolac-diazonaphthoquinone photoresist that evolved from similar materials developed for the manufacture of photoplates used in the printing industry in the early 1900's (23). The novolac matrix resin is a condensation polymer of a substituted phenol and formaldehyde that is rendered insoluble in aqueous base through addition of 10-20 wt% of a diazonaphthoquinone photoactive dissolution inhibitor (PAC). Upon irradiation, the PAC undergoes a Wolff rearrangement followed by hydrolysis to afford a base-soluble indene carboxylic acid. This reaction renders the exposed regions of the composite films soluble in aqueous base, and allows image formation. A schematic representation of the chemistry of this "solution inhibition" resist is shown in Figure 6.

The essential components of all conventional positive photoresists are the same. However, minor changes in the structure of the novolac and/or PAC will lead to changes in resist performance. Numerous investigations have shown that there is a delicate balance between PAC and resin structure, PAC weight fraction, developer, and process sequence (24,25). For example, a change in PAC structure from the 5-aryl sulfonate ester to the 4-aryl sulfonate analog optimizes the resist for exposure in the 310-365 nm region as opposed to the 350-450 nm range. Ring substitution in the "4" position leads to the appearance of a bleachable absorbance at ~315 and 385 nm as opposed to ~340, 400 and 430 nm for the 5-sulfonate material effecting shorter wavelength sensitivity for the resist. This is just one example of how optimization of resist sensitivity for a particular wavelength and exposure tool requires an understanding of the effect of substituents on the absorption characteristics of a material. In the case of conventional photoresist chemistry, Miller, et. al. (26,27), coupled such studies with semi-empirical calculations to facilitate the design of diazonaphthoquinone inhibitors for mid-UV applications.

While novolac-diazoquinone resists are sensitive, high resolution photoresists, they are essentially opaque to radiation below 300 nm. Use of these materials in this wavelength range would require extended exposure times to degrade the inhibitor near the resist-substrate interface and the resulting image quality would be poor. This has spurred considerable interest in developing new solution-inhibition resist chemistry for the deep-UV region. Willson and co-workers have reported on the utility of Medrum's acid derivatives (28,29) such as those shown in Figure 7, while workers at AT&T Bell Laboratories have employed nitrobenzyl ester photochemistry for deep-UV applications (Figure 8) (30).

Interest in solution inhibition resist systems is not limited to photoresist technology. Systems that are sensitive to electron-beam irradiation have also been of active interest. While conventional positive photoresists may be used for e-beam applications (31,32), they exhibit poor sensitivity and alternatives are desirable. Bowden, et al, at AT&T Bell Laboratories, developed a novel, novolac-poly(2-methyl-1-pentene sulfone) (PMPS) composite resist, NPR (Figure 9) (33,34). PMPS, which acts as a dissolution inhibitor for the novolac resin, undergoes spontaneous depolymerization upon irradiation (35). Subsequent vaporization facilitates aqueous base removal of the exposed regions. Resist systems based on this chemistry have also been reported by other workers (36,37).

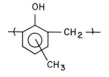

NOVOLAC STRUCTURE

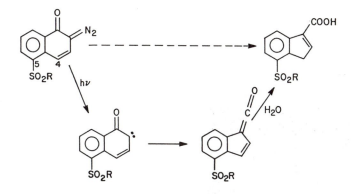

PAC PHOTOCHEMISTRY

Figure 6   Schematic representation of conventional positive photoresist chemistry.

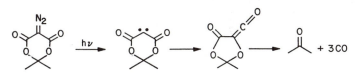

Figure 7   Schematic representation of Meldrum's Acid photochemistry.

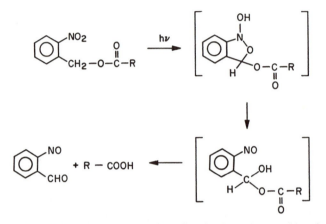

Figure 8   Schematic representation of o-nitrobenzyl ester photochemistry.

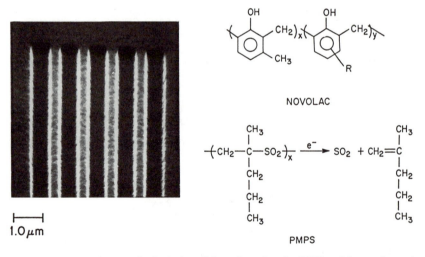

1.0 μm

NOVOLAC

PMPS

Figure 9   SEM micrograph depicting 0.5μm imaging in NPR with a schematic representation of NPR chemistry.

The classic positive resist that undergoes chain scission upon irradiation is poly(methyl methacrylate)(PMMA). PMMA was first reported by Hatzakis (38) as a high resolution electron-beam resist and is still considered to be one of the highest resolution materials available. The polymer degradation mechanism is believed to involve radiation induced cleavage of the side chain with resultant formation of a radical on the polymer backbone (Figure 10) (39). Subsequent β-scission leads to a reduction in polymer molecular weight and enhanced solubility of the exposed regions. Use of an appropriate developer allows selective removal of the irradiated areas.

While PMMA is an attractive material because of its resolution characteristics, its sensitivity to radiation induced degradation is low, and its dry-etching characteristics are poor. The e-beam and deep-UV exposure doses for PMMA are ~ 100 μC cm^{-2} (38) and >1 Jcm^{-2} (40), respectively. The fact that nanometer resolution is readily achieved in this material has, however, prompted many researchers to examine substituted systems in attempts to effect improved performance.

As mentioned above, the scission efficiency of a polymer may be given in terms of its G value. A higher value of G_s is an indicator of higher susceptibility to radiation induced degradation and improved resist sensitivity is expected. One of the first reports correlating methacrylate substitution pattern with ease of chain scission was published by Helbert et al (41). Here, it was predicted that copolymers of methyl methacrylate(MMA) with α-substituted chloro or cyano acrylates should effectively enhance the rate of polymer degradation. G_s values as high as 6.7 were determined for α-chloroacrylonitrile-MMA copolymers compared to 1.4 for the parent, PMMA. Methacrylonitrile-methyl α-chloroacrylate copolymers have since been prepared by Lai and coworkers that exhibit e-beam sensitivities ~ 5 times higher than PMMA (42,43). Generally, this body of work has shown that the efficiency of methacrylate chain scission can be enhanced by substitution of electronegative groups such as Cl or CN at the quaternary carbon (44). Incorporation of fluorine into the methacrylate chain is another method that has effectively enhanced PMMA sensitivity to radiation induced degradation. Polymers such as poly(2,2,3,4,4,4-hexafluorobutyl methacrylate) (45) and poly(2,2,2-trifluoroethyl-α-chloroacrylate) (46,47) are reported to have resist sensitivities less than 10 μC cm^{-2}.

Alternatively, introduction of bulky groups that provide steric hindrance, or structures that introduce strain into the polymer backbone may weaken the main chain. An example of the latter is the introduction of inter- and intra-molecular anhydride linkages into a basic methacrylate chain to enhance its e-beam sensitivity. The effectiveness of this chemistry was first demonstrated by Roberts, who prepared terpolymers of MMA, methacrylic acid (MAA) and methacryloyl chloride that upon prebaking generated intermolecular anhydride crosslinks (48,49). Radiation effects cleavage of these linkages thereby increasing the solubility of the polymer in the exposed regions. A terpolymer resist that utilizes intramolecular anhydride links was developed by workers at IBM. Optimization of the monomer ratios of this material (a terpolymer of MMA, MAA and methacrylic anhydride) (Figure 11) affords sensitivities of 5-10 μC cm^{-2} at 25 kV(50). Here, sensitivity enhancement occurs *via* both the incorporation of radiation sensitive anhydride linkages and excess strain in the polymer chain due to cyclization.

The deep-UV sensitivity of PMMA, >1 Jcm^{-2} (40), is also far from adequate to allow its use as a production scale UV resist. This is primarily due to the lack of

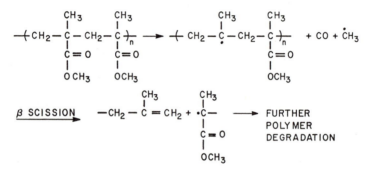

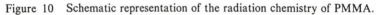

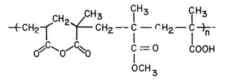

Figure 10   Schematic representation of the radiation chemistry of PMMA.

Figure 11   Schematic representation of poly(methyl methacrylate -co- methacrylic acid -co- methacrylic anhydride).

appreciable UV absorption at λ >230nm. Several investigators have effected improvements in the deep-UV sensitivity of PMMA *via* copolymerization of MMA with more absorbant, UV sensitive components. Examples include the copolymers of MMA with 3-oximino-2-butanone methacrylate (51) or indenone (52).

Another class of chain scission positive resists is the poly(olefin-sulfones). These materials are alternating copolymers of an olefin and sulfur dioxide, prepared by free radical solution polymerization. The relatively weak C-S bond, ~ 60 kcal/mole compared with ~ 80 kcal/mole for a carbon-carbon bond, is readily cleaved upon irradiation (G_s values for these polymers are typically ~ 10), and several sensitive resists have been developed based on this chemistry (53). One material that has been made commercially available is poly(butene-1-sulfone) (54).

Selected resists based on chemical amplification processes are capable of generating either positive or negative tone images upon development. These materials generally employ an onium salt initiator that dissociates upon irradiation to produce an acid that serves as a catalyst for a chain reaction of bond-forming or bond-breaking reactions. Use of this catalytic process effectively increases the quantum efficiency of the reaction well over the quantum yield for initial onium salt dissociation. The first example of a high resolution, positive resist employing onium salt chemistry was developed by Ito, et al (55,56). The thermally stable, acid labile, t-butyloxycarbonyl group was used to mask the hydroxy functionality of poly(vinylphenol). Mild heating of the masked polymer in the presence of an acid catalyst generated from the irradiation of an onium salt such as diphenyliodonium hexafluoroarsenate liberates the acidic hydroxyl group (Figure 12). This results in a large change in polarity in the exposed areas of the film and allows formation of either positive or negative images depending upon developer selection. Chemical amplification resists exhibit high sensitivity to deep-UV and electron-beam irradiation and may be sensitized to longer wavelengths through the addition of appropriate mid- and near-UV dyes.

NOVEL RESIST PROCESSES

The increased complexity of IC design has increased the demands placed on resist performance. New lithographic technologies and processing techniques will be required to achieve the necessary improvements in resolution and linewidth control (1). This has led many investigators to explore "non-traditional" areas of chemistry for resist applications. Since Taylor and Wolf (57) demonstrated that the incorporation of silicon into organic polymers can render them resistant to erosion in oxygen plasmas, there have been considerable research efforts aimed at the design of new polymers that incorporate silicon into the polymer structure (58). During oxygen etching, a 50-100 Å (59) thick protective layer of silicon oxide is formed on the polymer's surface that inhibits further attack of the film. This allows the use of these materials in a bilevel resist process such as that outlined in Figure 13. The device substrate is first coated with a thick layer of an organic polymer that effects planarization. Subsequently a radiation sensitive, etching resistant material is coated on its surface. Conventional processing allows pattern definition of the upper layer and that pattern is then transferred to the substrate by oxygen RIE techniques. The advantages of this process over conventional single layer processing include the following: alleviation of proximity effects during electron-beam irradiation, and elimination of substrate reflections and standing wave effects that occur during optical exposure.

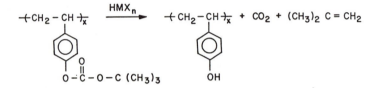

$$Ar_2I^+ \; MX_n^- \; \xrightarrow[RH]{h\nu} \; + HMX_n + ArI + Ar^\bullet + R^\bullet$$

$$Ar_3S^+ \; MX_n^- \; \xrightarrow[RH]{h\nu} \; + HMX_n + Ar_2S + Ar^\bullet + R^\bullet$$

$$MX_n = \text{e. g., } BF_4, \; PF_6, \; AsF_6, \; SbF_6$$

ONIUM SALT PHOTOCHEMISTRY

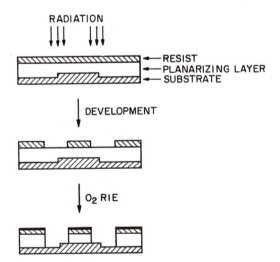

CATALYTIC DEPROTECTION OF MATRIX RESIN

Figure 12   Schematic representation of chemical amplification resist chemistry.

RADIATION

←RESIST
←PLANARIZING LAYER
←SUBSTRATE

DEVELOPMENT

O_2 RIE

Figure 13   Schematic representation of the bilevel resist process employing an oxygen reactive ion etching pattern transfer technique.

The oxygen RIE rate of organosilicon polymers is non-linear with respect to silicon content, and the incorporation of 10-15 wt% silicon leads to a significant reduction in etching rate (58). The oxygen etching rates of silicon containing polymers are related to the mass balance of silicon in the materials when high bombardment energy conditions, typical of RIE processing, are employed (60,61). Here, it appears that polymer structure has little if any effect on the etching behavior of a given resin, and silicon content is the only variable. Under low energy conditions, polymer structure plays an increasingly important role. The propensity of a given material to undergo radical induced degradation reactions appears particularly significant. Generally, the etching rate increases relative to predicted values with increasing ease of radical reaction. Further understanding of the relationship between polymer structure and composition, and etching conditions is required to achieve a viable bilevel resist system.

There are several problems encountered with silicon containing polymers that affect their lithographic properties. First, a decrease in T_g often accompanies silicon incorporation into a polymer. This may lead to dimensional instability of features during processing. Also, the hydrophobic nature of most useful silicon substituents may hinder the aqueous development of these resists. Careful selection of the polymer components can alleviate and/or eliminate these problems.

Several silicon-containing resist systems have recently been prepared and used in multilevel, RIE pattern transfer processes. Much of the chemistry of these systems closely resembles that of the conventional resists described above. Silyl substituted styrenes (59,62) and methacrylates (63,64) have been used in the preparation of both positive and negative acting resists such as those depicted in Figure 14. Additionally, silicon bearing groups have been incorporated into novolac resins that may then be used in conventional positive resist formulations (65,66). Copolymers of dimethylsiloxane, methylphenylsiloxane or methylvinylsiloxane are negative resists that are sensitive to both UV and electron-beam irradiation and exhibit high resolution imaging (67). However, it is necessary to keep the imaging layer thin (<3000 Å) to overcome problems such as image creep due to the low T_g of these materials. Chloromethylated phenylsiloxanes (68) and poly(silsequioxanes) (69,70) have been developed as high T_g alternatives to the original siloxane resists. Alternatively, block copolymerization of dimethylsiloxane and chloromethylstyrene combines the excellent RIE resistance of the siloxane unit with the radiation sensitivity of PCMS (71). Block copolymerization avoids problems associated with polymer phase separation and the flow characteristics of the resist are determined by the high T_g styrene unit. Polysilanes are a class of chain scission positive resists that contain silicon in the polymer backbone (72,73). These polymers have Si-Si bonds that undergo radical chain scission upon exposure to UV light (Figure 15).

Another attractive approach employed in the development of high resolution, high sensitivity resist systems is the elimination of wet-development in preference for a fully dry-developed resist system. Here, irradiation is used to effect a differential plasma etching rate in the resist. As first described by Taylor, et al, a guest monomer (M) is incorporated into a host polymer (P), and irradiation effects polymerization and some degree of grafting of the monomer (74). Appropriate selection of M allows plasma development of either positive or negative images (75,76). Vapor phase functionalization (77) is a new technique that is receiving increased attention as a method for combining the advantages of device planarization and solventless development. A schematic representation of the process is outlined in Figure 16.

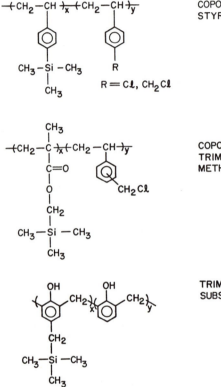

COPOLYMERS OF TRIMETHYLSILYL STYRENE

COPOLYMERS OF TRIMETHYLSILYLMETHYL METHACRYLATE

TRIMETHYLSILYLMETHYL SUBSTITUTED NOVOLAC RESINS

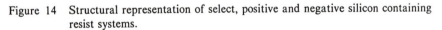

Figure 14   Structural representation of select, positive and negative silicon containing resist systems.

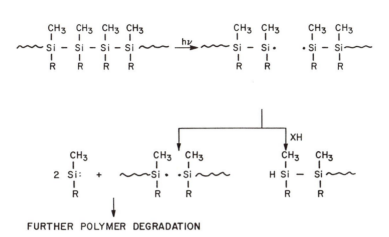

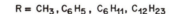

FURTHER POLYMER DEGRADATION

R = CH₃, C₆H₅, C₆H₁₁, C₁₂H₂₃

Figure 15   Schematic representation of the mechanism of polysilane chain scission.

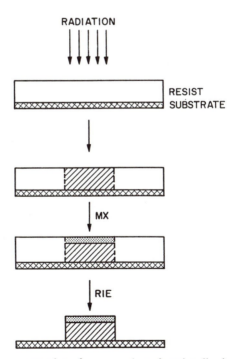

Figure 16   Schematic representation of a vapor phase functionalized resist process.

Inorganic reagents are generally used to selectively functionalize resist films after exposure to effect oxygen RIE resistance, and several examples have recently appeared in the literatures (78-80).

CONCLUSION

Polymer radiation chemistry is a key element of the electronics industry, in that polymer materials that undergo radiation induced changes in solubility are used to define the individual elements of integrated circuits. As the demands placed on these materials increases due to increased density, complexity and miniaturization of devices, new materials and chemistry will be required. This necessitates continued efforts to understand fundamental polymer radiation chemical processes, and continued development of new radiation sensitive materials that are applicable to VLSI Technology.

REFERENCES

[1]   "Introduction to Microlithography;" *ACS Symposium Series* **219**; Thompson, L. F., Willson, C. G. and Bowden, M. J., Eds.; American Chemical Society, Washington, D. C., 1983.

[2]   Pol, V., Bennewitz, J. H., Escher, G. C., Feldman, M., Firtion, V. A., et.al., *Proc. SPIE Opt. Microlithog. V*; 1986, **633**, 6.

[3]   Thompson, L. F., Bowden, M. J., in "Introduction to Microlithography;" *ACS Symposium Series,* **219**; Thompson, L. F.; Willson, C. G. and Bowden, M. J.; Eds. American Chemical Society, Washington, D. C. 1983, pp.15-85.

[4]   Novembre, A. E., Masakowski, L M., Hartney, M.A., *Poly. Eng. Sci.*, 1986, **26** (16), 1158.

[5]   Thompson, L. F., Kerwin, R. E., in "Ann. Rev. Mat. Sci.," Huggins, R. A., Bube, R. H. and Roberts, R. W., Eds., 1976, **6**, 267.

[6]   Feit, E. D., in "UV Curing: Science and Technology", Pappas, S. P., Ed.; Technology Marketing Corp., Stanford, CT, 1978, pp.230-247.

[7]   Iwayanagi, T., Kohashi, T., Nonagaki, S., Matsuzawa, T., Douta, K., Yanazawa, H., *IEEE Trans. Elec. Dev. Ed.*, 1981, **28** (11), 1306.

[8]   Bartelt, J. L., *J. Electrochemical Soc.,* 1975, **122** (4), 541.

[9]   Tan, Z. C., Petroupoulos, C. C., Rauner, F. J.; *J. Vac. Sci. Technol.;* 1981, **19**, (4), 1348.

[10]   Shu, J., Lee, W., Venable, L., Varnell, G.; *Poly. Eng. Sci.,* 1983, *23* (17), 1980.

[11]   Nonogaki, S., *Proc. Reg. Tech. Conf.* on "Photopolymers, Princ., Proc. and Mat.," Mid-Hudson Sect. SPE, Ellenville, N. Y., Nov. 8-10, 1982, p.1.

[12]   Hirai, T., Hatano, Y., Nonogaki, S.; *J. Electrochem. Soc.;* 1971, **118**, (4), 669.

[13]   Feit, E. D., Thompson, L. F., Heidenreich, R. D.; *ACS Div. of Org. Coat. and Plast. Chem. Preprints;* 1973, 383.

[14] Thompson, L. F., Feit, E. D., Heidenreich, R. D.; *Poly. Eng. Sci.* 1974, **14**(7), 529.

[15] Tabata, Y., Tagawa, S., Washio, M., in "Materials for Microlithography," Thompson, L. F., Willson, C. G. and Frechet, J. M. J., Eds.; ACS *Symposium Series*, **266**, American Chemical Society, Washington, D. C., 1984, pp.151-163.

[16] Taylor, G. N., Wolf, T. M., *J. Electrochem. Soc.*, 1980, **127**, 2665.

[17] Thompson, L. F., Yau, L., Doerries, E. M., *J. Electrochem. Soc.*, 1979, **126**(19), 1703.

[18] Novembre, A. E., Bowden, M. J.; *Poly. Eng. Sci.*, 1983, **23**(17), 975.

[19] Feit, E. D., Thompson, L. F., Wilkins, C. W., Jr., Wurtz, M. E., Doerries, E. M., Stillwagon, L. E.; *J. Vac. Sci. Technol.*, 1979, **16**, (6), 1997.

[20] Imamura, S., *J. Electrochem. Soc.*, 1979, **126** (9), 1268.

[21] Hartney, M. A., Tarascon, R. G., Novembre, A. E., *J. Vac. Sci. Technol.; B.* 1985, **3**, 360.

[22] Choong, H. S., Kahn, F. J.; *J. Vac. Sci. Technol.* 1981, **19**, (14), 1121.

[23] Kosar, J.; "Light Sensitive Systems," John Wiley and Sons, New York, N. Y., 1965, p. 194.

[24] Hanabata, M. Furuta, A. Uemura, Y.; *Proc. SPIE Advances in Resist Technology*; 1986, **681**, 76.

[25] Trefonas, P., III, Daniels, B. K.; *Proc. SPIE Advances in Resist Technology and Processing IV*; 1987, **771**, 194.

[26] Miller, R. D., Willson, C. G., McKean, D. R., Tompkins, T., Clecak, N., Michl, J. Downing, J.; *Proc. Reg. Tech. Conf.* on "Photopolymers: Princ., Proc., and Mat., "Mid-Hudson Sect. SPE, Ellenville, N. Y., Nov. 8-10, 1982, p.111.

[27] Miller, R. D., Willson, C. G., McKean, D. R., Tompkins, T., Clecak, N., Michl, J., Downing, J.; *Org. Coat. and Plast. Chem. Preprints*, 1983, **48**, 54.

[28] Willson, C. G., Clecak, N., Grant, B. D., Twieg, R. J.; *Electrochem. Soc. Preprints*, St. Louis, 1980, p. 696.

[29] Willson, C. G., Miller, R. D., McKean, D. R.; *Proc. SPIE Advances in Resist Technol. and Proc. IV*, 1987, **771**, 2.

[30] Reichmanis, E., Wilkins, C. W., Jr., Chandross, E. A.; *J. Vac. Sci. Technology*; 1981, **19**(4), 1338.

[31] Shaw, J. M., Hatzakis, M.; *J. Electrochem. Soc.; 1979,* **126** 2026.

[32] Shaw, J. M., Hatzakis, M.; *IEEE Trans. Electron. Dev.*, 1978, **25**, 4.

[33] Bowden, M. J., Thompson, L. F., Fahrenholtz, S. R., Doerries, E. M.; *J. Electrochem. Soc.*, 1981, **128**, 1304.

[34] Tarascon, R. G., Frackoviak, J., Reichmanis, E., Thompson, L. F.; *Proc. SPIE Advances in Resist Technol. and Proc. IV* , 1987, **771**, 54.

[35] Bowden, M. J., Allara, D. L. Vroom, W. I., Frackoviak, J., Kelley, L. C., Falcone, D. R.; in "Polymers in Electronics," Davidson, T., Ed.; *ACS*

Symposium Series, **241**, American Chemical Society, Washington, D. C., 1984, pp. 135-52.

[36] Chang, Y. Y., Grant, B. D., Pederson, L. A., Willson, C. G., US patent 4, 398, 001, 1983.

[37] Shiraishi, H., Isobe, A., Murai, F., Nonogaki, S., in "Polymers in Electronics," Davidson, T., Ed; *ACS Symposium Series*, **242**, American Chemical Society, Washington, D. C., 1984, pp. 167-76.

[38] Hatzakis, M.; *J. Electrochem. Soc.*; 1969, **116**, 1033.

[39] Ranby, B., Rabek, J. F.; "Photodegradation, Photooxidation and Photostabilization of Polymers," John Wiley and Sons, New York, N. Y., 1975, p. 573.

[40] Lin, B. J., *J. Vac. Sci. Technol.*, 1975, **12**, 1317.

[41] Helbert, J. N., Wagner, B. E., Caplan, J. P., Poindexter, E. H.; *J. Appl. Poly. Sci.*; 1975, **19**, 1201.

[42] Lai, J. H., Helbert, J. N., Cook, C. F., Pittman, C. U., Jr.; *J. Vac. Sci. Technol.*; 1979, **16**, 1992-95.

[43] Lai, J. H., *13th Annual SPSE Conf.*, May 18-22, 1986, Minneapolis, Minn.

[44] Helbert, J. N., Chen., C. Y., Pittman, C. U., Jr., Hagnauer, G. L.; *Macromolecules*; 1978, **11**, 1104.

[45] Kakuchi, M., Sugawara, S., Murase, K., Matsugoma, K.; *J. Electrochem. Soc.*; 1977, **124**, 1648.

[46] Tada, T.; *J. Electrochem. Soc.*; 1979, **126**, 1829.

[47] Tada, T.; *J. Electrochem. Soc.*; 1983, **130**, 912.

[48] Roberts, E. D.; *ACS Div. Org. Coat. and Plast. Chem. Preprints* 1973, **33** (1), 359.

[49] Roberts, E. D.; *ACS Div. Org. Coat. and Plast. Chem. Preprints* 1977, **37** (2), 36.

[50] Moreau, W., Merritt, D., Moyer, W., Hatzakis, M., Johnson, D., Pederson, L.; *J. Vac. Sci. Technol.*; 1979, **16** (6), 1989.

[51] Reichmanis, E., Wilkins, C. W., Jr.; in "Polymer Materials for Electronic Applications," Feit, E. D., Wilkins, C. W., Jr., Eds.; *ACS Symposium Series*, **184**, American Chemical Society, Washington, D. C., 1982, pp. 29-43.

[52] Hartless, R. L., Chandross, E. A.; *J. Vac. Sci. Technol.*; 1981, **19**, 1333.

[53] Bowden, M. J., Thompson, L. F.; *Solid State Technol.*; 1979, **22**, 72.

[54] Bowden, M. J., Thompson, L. F., Ballantyne, J. P.; *J. Vac. Sci. Technol.*; 1975, **12**, 1294.

[55] Ito, H., Willson, C. G., Frechet, J. M. J., Farrall, M. J., Eichler, E.; *Macromolecules*; 1983, **16**, 570.

[56] Ito, H., Willson, C. G.; in "Polymers in Electronics," Davidson, T., Ed.; *ACS Symposium Series*; **242**, 1984, pp. 11-23.

[57] Taylor, G. N., Wolf, T. M.; *Poly. Eng. Sci.*; 1980, **20**, 1087.

[58] Reichmanis, E., Smolinsky, G., Wilkins, C. W., Jr.; *Solid State Technol*; 1985, **28** (8) 130.

[59] Suzuki, M., Saigo, K., Gokan, H., Ohnishi, Y.; *J. Electrochem. Soc.*, 1983, **130**, 1962.

[60] Watanabe, F., Ohnishi, Y.; *J. Vac. Sci. Technol. B*; 1986, **4** (1), 422.

[61] Jurgensen, C. W., Shugard, A., Dudash, N., Reichmanis, E., Vasile, M. J.; *Proc. SPIE Advances in Resist Technol. and Proc. V*; 1988, **920**.

[62] MacDonald, S. A., Steinmann, A. S., Ito, H., Hatzakis, M., Lee, W., Hiraoka, H., Willson, C. G.; *ACS Proc. Polym. Mat. Sci. and Eng.*; 1983, **49**, 104.

[63] Novembre, A. E., Reichmanis, E., Davis, M. A.; *Proc. SPIE Advances in Resist Technol. and Proc. III*; 1986, **631**, 14.

[64] Reichmanis, E., Smolinsky, G; *J. Electrochem. Soc.*; 1985, **132**, 1178.

[65] Tarascon, R. G., Shugard, A., Reichmanis, E., *Proc. SPIE Advances in Resist Technol. and Proc. III*; 1986, **631**, 40.

[66] Saotome, Y., Gokan, H., Saigo, K., Sazuki, M., Ohnishi, Y.; *J. Electrochem. Soc.*; 1985, **132**, 909.

[67] Hatzakis, M., Paraszczak, J., Shaw, J. M., in "Microcircuit Engineering 81" Osgenbrug, A., Ed., Swiss Fed. Inst. Technol.; Lausanne, 1981, pp. 386-96.

[68] Tanaka, A., Morita, M., Imamura, S., Tamamura, T., Kogure, O.; *Japan J. Appl. Phys.*; 1983, **22**(10), 2659.

[69] Morita, M., Tanaka, A., Onoge, K.;*J. Vac. Sci. Technol. B.*; 1986, **4** (1), 414.

[70] Brault, R. G., Kubena, R. L., Metzger, R. A.; *Proc. SPIE Advances in Resist Technol.*; 1985, **539**, 70.

[71] Hartney, M. A., Novembre, A. E., Bates, F. S. *J. Vac. Sci. Technol. B;* 1985, **3**(5), 1346.

[72] Miller, R. D., Hofer, D. C., Willson, C. G., West, R., Trefonas, P. T. III; in "Materials for Microlithography," Thompson, L. F., Willson, C. G., Frechet, J. M. J., Eds.; *ACS Symposium Series*; **266**, American Chemical Society, Washington, D. C., 1984, pp. 293-310.

[73] Zeigler, J. M., Harrah, L. A., Johnson, A. W.; *Proc. SPIE Advances in Resist Technol. and Proc. II*; 1985, **539**, 166.

[74] Taylor, G. N.;*Solid State Technol.*; 1980, **23**(5), 73.

[75] Taylor, G. N., Wolf, T. M.;*J. Electrochem. Soc.*, 1980, **127**(12), 2665.

[76] Taylor, G. N., Wolf, T. M., Moran, J. M.;*J. Vac. Sci. Technol.*, 1981, **19**(4), 872.

[77] Stillwagon, L. E., Silverman, P. J., Taylor, G. N.; *Proc. Reg. Tech. Conf.* Photopolymers: Princ., Proc. and Mat.," Mid-Hudson Sect., SPE, Ellenville, N. Y., Oct. 28-30, 1985, pp. 87-103.

[78] MacDonald, S. A., Ito, H., Hiraoka, H., Willson, C. G.; ibid., pp. 177-96.

[79] Roland, B., Lombaerts, R., *Proc. SPIE Advances in Resist Technol. and Proc. IV.*; 1987, **771**, 69.

[80] Schellekens, J. P. W., Visser, R. J., Reuhman-Huisken, M. E., Van Yzendoorn, L. J.; ibid, 111.

RECEIVED August 26, 1988

Chapter 10

Imaging Processes Based on Side-Chain Modification of Polymers

Synthesis and Study of Allylic and Benzylic Ethers Derived from Poly(vinylphenols)

Jean M. J. Fréchet[1,2], Eva Eichler[2], Sylvie Gauthier[2], Bogustav Kryczka[2], and C. G. Willson[3]

[1]Baker Laboratory, Department of Chemistry, Cornell University, Ithaca, NY 14853–1301
[2]Department of Chemistry, University of Ottawa, Ontario K1N 9B4, Canada
[3]Almaden Research Center, IBM Corporation, San Jose, CA 95120

The first use of ether protecting groups in the design of imaging systems based on substituted poly(hydroxy-styrenes) is reported. Polymers containing 4-(2-cy-clohexenyloxy) or 4-(1-phenylethyloxy) derivatives of 4-vinylphenol or 3,5-dimethyl-4-vinylphenol have been prepared from the corresponding monomers. Due to their design, which allows for facile elimination or rearrangement reactions, the ether protecting groups can be removed easily by acidolysis, or thermolysis, or a combination thereof. In some instances, the protecting groups can be split quantitatively from the polymers, while in others a thermal Claisen rearran-gement or an acid-catalyzed alkylation occur with the formation of some alkylated phenolic moieties. Appli-cation of the design to imaging systems is achieved through the use of triarylsulfonium salts as photo-chemical triggers. Exposure of films of poly[4-(2-cyclohexenyloxy)-3,5-dimethyl-styrene] containing some of the onium salt to irradiation at 254 nm results in the formation of acid in the exposed areas which cata-lyzes the polymer deprotection and allows for the development of images in either positive or negative mode through a differential dissolution process.

The radiation induced side-chain modification of polymers containing pendant phenyl ester groups has been the object of several studies as some undergo a photo-Fries rearrangement (1) while others such as

0097–6156/89/0381–0155$06.00/0
© 1989 American Chemical Society

the formate esters of poly (p-hydroxystyrene) are cleaved cleanly in a process which can be used for the design of a new type of dual-tone imaging systems (2). Numerous other polymers with ester or carbonate pendant groups have also found applications as resist materials in microlithography (3-6). Thus, two families of new and potentially useful styrene based imaging systems have emerged. One is based on functional derivatives of poly (p-hydroxystyrene) (3,4) or similar substituted x-methylstyrene (5), while the other is based on functional derivatives of poly (p-vinylbenzoic acid) (6). In both cases imaging rests on the ability of the polymer to undergo photoinitiated removal of the phenolic or carboxylic acid protecting group, usually esters or carbonates, in a process which results in a drastic change of polarity for those areas of the polymer films which have been exposed to irradiation (3,7).

Figure 1 outlines the principle of such imaging which is based on the well-known principle of differential dissolution. The starting ester or carbonate polymers are soluble in ordinary organic solvents such as aromatics and halogenated hydrocarbons, while the free phenolic or carboxylic acid polymers obtained as a result of radia-tion-induced processes are soluble in aqueous base. Thus, develop-ment of an imagewise exposed coating of such polymer may be accom-plished either through removal of the exposed areas of the coating using aqueous base to afford a positive-tone image, or through remo-val of the unexposed areas to afford a negative-tone image. Such imaging systems are noteworthy as they offer great versatility through their dual mode of development, as well as excellent resolution due to the absence of swelling in the development process (7). This lack of swelling is primarily the consequence of the large change in polarity which results from cleavage of the side-chain phenolic or carboxylic protecting groups.

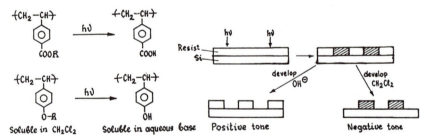

Figure 1. Imaging through changes in polymer side-chain polarity.

Results and Discussion

While carbonates and esters derived from functionalized polystyrenes have received much attention, ethers have mostly been ignored as the cleavage of ether groups is generally considered to be a difficult reaction. For example, very harsh reaction conditions involving reagents such as BBr_3 are required to cleave phenylalkyl ethers (8). A notable exception is the t-butyl ether group which is readily cleaved in acidic medium and has been tested in the protection of some amino acids such as serine or 4-hydroxyproline (9).

In the course of a broad model study on the use of certain carbonates, such as I in Scheme I, as easily removable protecting groups for alcohols or phenols (10) it was observed that significant yields of ethers having structure III may be formed during the acid-catalyzed thermolytic cleavage of the carbonate functionalities of I (11). The overall process amounts to the transformation of a carbonate into an ether with extrusion of carbon dioxide. In fact, a simple cationic mechanism may be invoked to explain this reaction as is shown in Scheme I. The reaction is facilitated by the relatively facile formation of the stabilized benzylic carbocation II which can readily recombine with the nucleophilic alcohol ROH which is also formed in the first stage of the reaction. An alternate pathway for reaction of the benzylic carbocation, also shown in Scheme I, involves elimination of a proton with formation of the conjugated alkene: styrene. The latter reaction may be used in other ways as will be described more fully elsewhere (11a). Subsequent experimentation with ethers IIIa or IIIb showed that they were also susceptible to acid-catalyzed thermolysis which occurred with evolution of ethanol, or p-nitrophenol, and styrene.

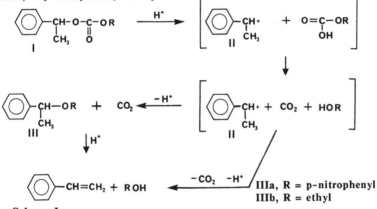

Scheme I

In view of the proposed mechanism for this cleavage reaction, it was reasonable to expect that certain allylic ethers for which an elimination pathway existed (Scheme II) would also be susceptible to cleavage under thermolysis, acidolysis or a combination of both.

Scheme II

<u>Synthesis of the Polymers Containing Reactive Ether Pendant Groups.</u>
We have previously described the synthesis of high purity poly(p-hydroxystyrene) free from deleterious oxidized species by polymerization of 4-t-butyloxycarbonyloxy-styrene followed by the removal of

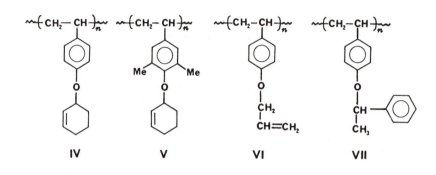

IV   V   VI   VII

the t-BOC protecting group (<u>4</u>). A less pure but nevertheless satisfactory grade of poly(p-hydroxystyrene) can be obtained commercially
from Maruzen Oil Co. (Resin PHM). Both types of poly(p-hydroxystyrene) were used satisfactorily as starting materials for the
preparation of polymer VI which contains pendant allyl ether groups.
The same simple nucleophilic displacement reaction using poly(p-
hydroxystyrene) as the source of nucleophile can also be used to
prepare allylic polymers such as IV or V. Alternately, polymers IV,
V and VII have also been prepared by polymerization of the corresponding monomers as it is possible to avoid any involvement of the
cyclohexenyloxy substituent in the radical polymerization of the
styrenic moieties. Scheme III shows the preparation of 3,5-dime-
thyl-4(2-cyclohexenyloxy)styrene. The starting material in this
reaction is 3,5-dimethyl-4-hydroxy-benzaldehyde which is itself
obtained by formylation of 2,6-dimethylphenol. A Williamson ether
synthesis with 3-bromocyclohexene under phase transfer conditions
affords the allylic ether which can then be used in a high yield
Wittig reaction with methyl triphenylphosphonium bromide to afford
the styrenic monomer (Scheme III). The allylic ether monomer
derived from 3,5-dimethyl-4-(2-cyclohexenyloxy)-benzaldehyde was
prepared similarly by a Wittig reaction, while the benzylic mono-
meric precursor of VII was obtained by reaction of p-hydroxybenzaldehyde with 1-bromo-1-phenyl-ethane followed by Wittig methylena-
tion. The monomers were then polymerized under free-radical conditions using AIBN as initiator in toluene, to afford the desired
polymers IV, V and VII with pendant allyl or benzyl ether groups.

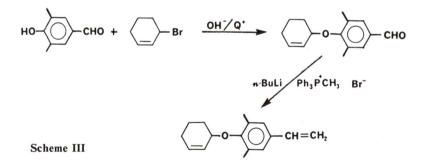

Scheme III

Thermolysis of the Polymers.   The thermolytic behavior of polymers IV and V was studied first by thermogravimetry. The thermogravimetric analysis (TGA) of IV is shown in Figure 2; it indicates clearly that a cleavage reaction occurs near 230°C. An analysis of the volatile product of the thermolysis by a combination of thermolysis, gas chromatography, and mass spectrometry, confirmed that the expected 1,3-cyclohexadiene elimination product had indeed been produced. However, careful analysis of the TGA data showed that only a fraction of the original allylic groups had been removed during thermolysis. This is not too unexpected in view of the known propensity of phenyl allyl ethers to undergo a thermal Claisen rearrangement. The overall thermolytic behavior of polymer IV is represented in Scheme IV. A similar study with poly (p-2-propenyloxystyrene) VI shows that no such thermal cleavage reaction occurs as the 2-propenyloxy group has no simple elimination pathway which would allow the formation of a conjugated diene. The thermogravimetric analysis of polymer VI only shows that the thermal degradation of VI resembles that of poly(p-hydroxystyrene) itself.

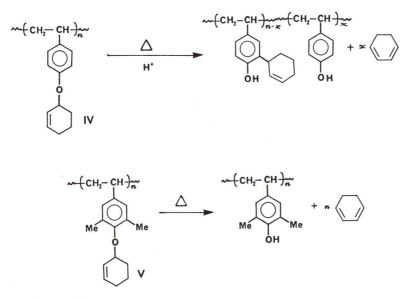

Scheme IV

In contrast, and as seen on Figure 3, polymer V undergoes complete thermolytic cleavage of its pendant ether groups when heated to 250°C as the Claisen rearrangement, which might normally compete with the desired cleavage reaction, is prevented by the presence of alkyl substituents in the ortho and para positions of the styrenic aromatic rings of the starting polymer. Here again monitoring of the thermolysis by gas chromatography-mass spectrometry shows that 1,3-cyclohexadiene is evolved cleanly at temperatures as low as 234°C. Within experimental errors, it is observed that the weight loss in this reaction corresponds to that expected for complete

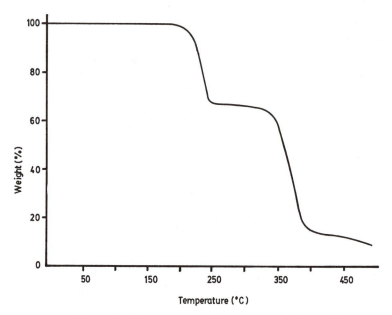

Figure 2. Thermogravimetric analysis of polymer IV.

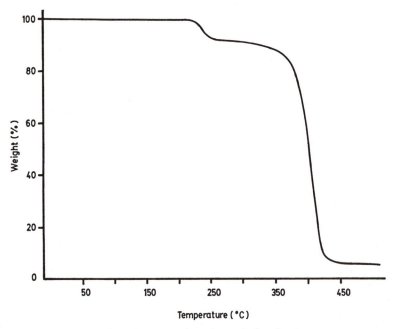

Figure 3. Thermogravimetric analysis of polymer V.

removal of the 2-cyclohexenyl protecting groups as 1,3-cyclohexa-diene. The overall thermolytic behavior of polymer IV is also shown in Scheme IV. Although it had been expected that polymer VII which contained reactive pendant benzylic ether groups would also undergo a relatively facile thermolytic cleavage, TGA experiments showed clearly that this was not the case. The TGA curve which is given in Figure 4, together with the derivative curve, indicate that a first decomposition occurs near 320°C instead of the much lower temperature which had been observed with the allylic ether polymers IV and V. This first decomposition is soon followed by a second decomposition which corresponds to the complete breakdown of the polymer. Though the initial weight loss centered near 320°C roughly corresponds to the loss of styrene, it cannot be said that this step in the TGA curve corresponds only to the thermolysis of the ether pendant groups as the liberated styrene may also polymerize to oligomeric species which would decompose subsequently. A comparison of the thermolytic data which that obtained for pure poly(p-hydroxy-styrene) (Figure 5) shows that in the absence of phenolic protecting groups the thermolysis proceeds to leave a larger amount of residue above 350°C, likely due to oxidative coupling reactions.

Imaging Experiments. Most imaging experiments were carried out using polymers IV and V as both VI and VII were less suitable in the context of this study. The imaging strategy which was used was an extension of earlier approaches (3,6), whereby the polymers are used in combination with a photoactive onium salt which liberates strong acid upon irradiation. It was expected that acid would catalyze the thermolytic cleavage of the allylic ethers as it does in the case of model compounds. In addition, the possibility remained that acid would also catalyze the Claisen rearrangement (Scheme IV) of polymer IV; however, this should not adversely affect the overall imaging process as both modes of reaction, cleavage or rearrangement, would produce free phenolic groups and would therefore be suitable for imaging purposes.

UV spectra of both polymers IV and V taken from ca. 1μm thick films showed that the each of the polymers had absorbencies of less than 0.30 per micron of film thickness at 254 nm as shown in Figure 6 for polymer V. This allows their use with triarylsulfonium salts, which generally absorb in the deep UV. In addition, it is known that poly(4-hydroxystyrene) does not absorb strongly in the deep UV, while poly(3,5-dimethyl-4-hydroxystyrene) also shows no strong absorption band near 254 nm (Figure 6).

In view of these findings, all samples for imaging were prepared as follows. The polymer solutions (10-20wt % of polymer) were prepared using cyclohexanone as solvent and triphenylsulfonium hexafluoro-antimonate (10-12% by weight with respect to polymer) was used as the onium salt. The solutions were then spin-coated onto 5" silicon wafers and imaging was carried out using a Perkin-Elmer 500 projection printing tool fitted with appropriate filters. Exposure to UV radiation at 254 nm was followed by baking at 105°C, which caused a visible image to appear. Sensitivity measurements showed that both polymers had sensitivities of better than 10 mJ/cm^2 and good contrast. Polymer V behaved as a dual-tone resist affording posi-

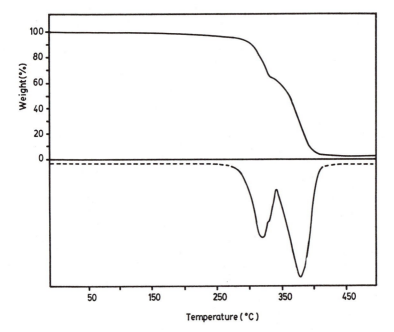

Figure 4. Thermogravimetric analysis of polymer VII (upper curve) with derivative (lower curve) showing two distinct thermal degradations.

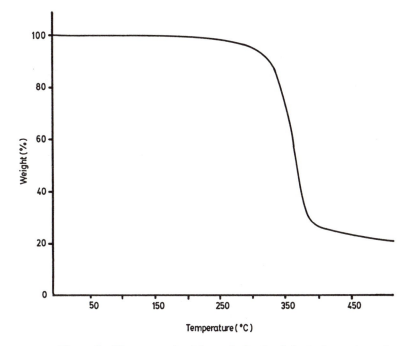

Figure 5. Thermogravimetric analysis of poly(p-hydroxystyrene).

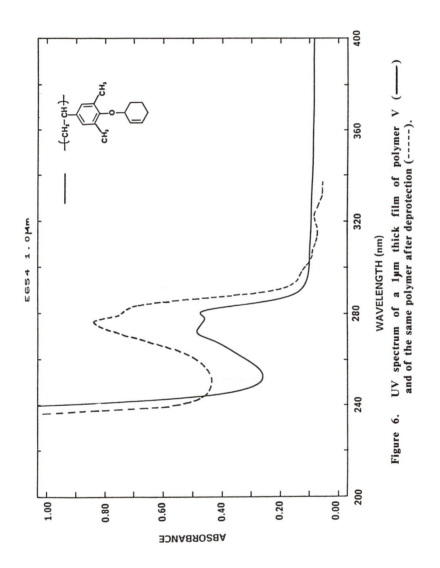

Figure 6. UV spectrum of a 1μm thick film of polymer V (———)
and of the same polymer after deprotection (- - - -).

tive or negative images of the mask depending on the solvent chosen for development. Dissolution of the exposed areas with dilute ethanolic potassium hydroxide gave a positive-tone image while development with toluene-hexane mixtures gave the complementary negative-tone image. It should be noted that the large change in polarity which occurs as the ether groups are thermolyzed in the exposed areas contributes greatly to the overall quality of the imaging process as no swelling is observed during development. The imaging process with polymer V is therefore similar to that which is outlined schematically on Figure 1; a micrograph of a positive image produced using this process is shown in Figure 7. In addition to monitoring the overall process through imaging results, the removal of pendant ether groups can also be followed by FT-infrared spectroscopy using a film of the polymer containing 10wt% of the onium salt cast on a sodium chloride disc. Figure 8 shows the infrared spectrum of a film of the starting polymer (upper curve) as well as the spectrum of the polymer after exposure and baking (lower curve).

Imaging of polymer IV proved to be less attractive probably due to the occurrence of some Claisen rearrangement and of other side reactions involving the rearranged polymer. Formation of an insoluble skin was observed and development with polar solvents such as 2-propanol, aqueous or alcoholic base proved ineffective. Similarly, attempted development with ketones such as cyclohexanone or 4-methylpentanone did not provide access to a fully developed positive image but caused swelling to occur.

Conclusion.

The use of simple mechanistic considerations in establishing the basic design features of imaging systems is once again demonstrated with the application of novel ether chemistry to photoinitiated side-chain modification. Imaging processes based on acid catalyzed thermolytic cleavage of appropriately chosen ethers of phenolic polymers can afford materials which operate both in positive and negative-tone mode.
While this work has focused on certain allylic and benzylic ether derivatives of poly(4-hydroxystyrene) it is possible to extend the concept to the corresponding t-butoxy substituted polymers although the preparation of such polymers is relatively more difficult than in the present case.

Experimental.

Thermogravimetric and Differential Scanning Calorimetry experiments were performed on Mettler TGA and DSC instruments. Mass Spectra were measured on a VG-7070E double-focusing mass spectrometer. GPC analysis were performed by comparison to polystyrene standards on a Waters model 150 gel permeation chromatograph equipped with five microstyragel columns using THF as the mobile phase. Preparative HPLC separations were carried out using a Waters 500 HPLC equipped with two 500g silica gel columns. NMR measurements were obtained on Varian XL-300 or CFT-80 instruments with $CDCl_3$ as solvent and TMS as internal standard. Infrared analyses were done on a Nicolet 10-DX FT-IR instrument using KBr pellets.

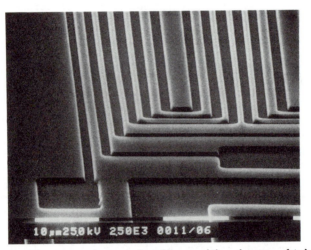

Figure 7.   Electron micrograph of a positive image obtained by the side-chain deprotection process.

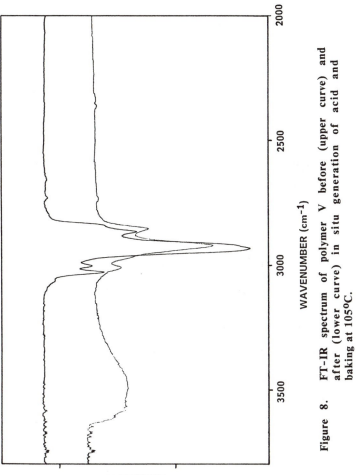

Figure 8. FT-IR spectrum of polymer V before (upper curve) and after (lower curve) in situ generation of acid and baking at 105°C.

Preparation of 4-(2-cyclohexenyloxy)-benzaldehyde. A solution of 16.1g (0.100 mole) of 3-bromocyclohexene, 3.2g (0.010 mole) tetra-butylammonium bromide and 12.2g (0.100 mole) 4-hydroxybenzaldehyde in 100ml dichloromethane was treated with a solution of 50g KOH in 50ml water with stirring. Stirring was continued for 24 hours with intermittent chromatographic monitoring. Additional dichloromethane and water were then added to facilitate separation of the organic and aqueous layers. The organic layer was then washed four times with 50ml water then dried over magnesium sulfate. After evapo-ration of the solvent and purification by preparative HPLC using 10% ethyl acetate in hexane as eluent, 17.4g of the pure product was obtained (86% yield). Spectroscopic analyses of the product confirm that it is the expected 4-(2-cyclohexanyloxy)-benzaldehyde. Elemental analysis. Calculated for $C_{13}H_{14}O_2$ (MW 202.24): C, 77.20; H, 6.98; O, 15.82. Found: C, 76.87; H, 7.07.

Preparation of 4-(2-cyclohexenyloxy)-styrene. A stirred mixture of 34.36g (0.096 mole) methyltriphenylphosphonium bromide and 10.75g (0.096 mole) potassium t-butoxide in 200ml dry THF is treated drop-wise with a solution of 16.16g (0.080 mole) of 4-(2-cyclohexenyl)-benzaldehyde in 30ml THF under inert atmosphere. Once the addition of aldehyde was completed, the mixture was stirred at room tempera-ture for another 2 hours. Ether and water were then added to the reaction mixture until clearly separated phases were obtained with no solid residue. The organic layer was separated and washed three times with water, dried over magnesium sulfate and evaporated. The resulting semi-solid was triturated in 10% ethyl acetate-hexane mixture to remove most of the triphenylphosphine and the evaporated extract was purified by preparative HPLC using hexane as eluent. This afforded 9.35g (58%) of the pure monomer, which was fully characterized by ^{1}H and ^{13}C-NMR as well as mass spectrometry.

Preparation of 4-(1-phenylethyloxy)benzaldehyde. This benzylic ether was prepared from p-hydroxybenzaldehyde (8.05 or 0.066 mole), and 1-bromo-ethylbenzene (12.1g or 0.065 mole) under phase transfer conditions as described above for 4-(2-cyclohexenyloxy)-benzalde-hyde. After purification by preparative HPLC 13.9g (93%) of pure product was obtained. Elemental analysis. Calculated for $C_{15}H_{14}O_2$ (MW = 226.26): C, 79.62; H, 6.24; O, 14.14. Found: C, 79.36; H, 6.39.

Preparation of 4-(1-phenylethyloxy)-styrene. This benzylic ether of p-hydroxystyrene was prepared by a Wittig reaction on the precursor aldehyde as described above. The final product was obtained in 73% yield after purification by preparative HPLC using 5% ethyl acetate in hexane as eluent. The product had analytical characteristics in agreement with the proposed structure.

Preparation of 3,5-dimethyl-4-hydroxybenzaldehyde. A mixture of 40.6g (0.33 mole) of 2-6-dimethylphenol and 46.2g (0.33 mole) of hexamethylene tetramine was covered with 500ml trifluoroacetic acid and the reaction mixture was refluxed overnight. After evaporation of the trifluoroacetic acid the residue was poured into 3L water and neutralized slowly with sodium carbonate. A brown oily solid was formed and extracted 3 times with ether. The combined ether extracts were then washed with a small amount of 1N HCl then with water. After drying over magnesium sulfate and filtration through silica gel, the crude product (over 97% pure by HPLC) was obtained (39.6g or 80% yield).

Preparation of 3.5-dimethyl-4(2-cyclohexenyloxy)benzaldehyde.
A solution of 22.2g (0.137 mole) 3-bromocyclohexene, 20.8g (0.137 mole) 3.5-dimethyl-4-hydroxybenzaldehyde and 4.45g tetrabutylammonium bromide in 200ml dichloromethene was stirred overnight with a solution of 75g KOH in 75ml water. Work-up was accomplished by adding 200ml ether and enough water to ensure complete separation of the aqueous and organic layers. The organic layer was collected and washed 5 times with water adding small portions of ether as necessary to keep the phases separated. After drying over magnesium sulfate and filtering through silica gel, 28.5g of crude thick oily product was obtained. Purification by preparative HPLC using 5% ethyl acetate in hexane as eluent afforded 20.6g (67.5%) of pure product. Elemental Analysis; calculated for $C_{15}H_{18}O_2$ (230.29) C, 78.22; H, 7.88. Found: C, 78.43; H, 7.61.

Preparation of 3.5-dimethyl-4-(2-cyclohexenyloxy)styrene. A well stirred mixture of 36.52g (0.102 mole) methyltriphenylphosphonium bromide and 11.42g (0.102 mole) potassium t-butoxide in 200ml dry tetrahydrofuran was treated dropwise with a solution of 19.55g (0.085 mole) 3.5-dimethyl-4-(2-cyclohexenyloxy)benzaldehyde in 50ml dry THF under nitrogen atmosphere. Following complete addition of the aldehyde, the mixture was stirred for a further 2 hours and enough ether and water were added to obtain two separate phases with no solid residue. The ether phase was collected while the aqueous phase was extracted with ether and the combined ether extracts were washed 3 times with water. After drying over magnesium sulfate, the ether solution was filtered over silica gel and evaporated to a semi-solid. Triphenylphosphine oxide was removed from the product by extraction with hexane. Chromatographic purification of the product over silica gel using 20% ethyl acetate in hexane as eluent afforded 18.6g (96%) of the pure product. Mass spectral and NMR analyses confirmed the identity of the product.
Elemental Analysis. Calculated for $C_{16}H_{20}O$ (228.32): C, 84.16; H, 8.83; O, 7.01. Found: C, 84.58; H, 8.68.

Preparation of Polymer IV. A solution of 8.0g (0.040 mole) 4-(2-cyclohexenyloxy)-styrene and 0.080g AIBN in 16g toluene was heated to 75°C under inert atmosphere. After 48 hours the resulting polymer was diluted with a minimum of toluene and precipitated into 3L of petroleum ether. The recovered polymer (6.1g, 76% yield) had spectral properties in agreement with the proposed structure.
Elemental analysis. Calculated for $(C_{14}H_{16}O)_n$: C, 83.96; H, 8.05; O, 7.99. Found: C, 84.40; H, 8.31. GPC. Mn: 15,400; MW: 27,400.

Preparation of Polymer V. A solution of 8g of monomer in 80g benzene was heated in an oil bath at 75°C under nitrogen atmosphere. After addition of 0.050g AIBN, the mixture was stirred then left at 75°C for 48 hours. After partial evaporation of the solvent, the polymer was precipitated in methanol, re-dissolved in THF and reprecipitated in methanol to afford 6.48g (81%) of the desired polymer. NMR analysis was consistent with the proposed structure. Elemental analysis. Calculated for $(C_{16}H_{20}O)_n$: C, 84.16; H, 8.83; O, 7.01. Found: C, 83.95; H, 8.89. GPC. Mn: 54,200; Mw: 97,500.

Preparation of Polymer VII. A solution of 8.0g (36 moles) 4-(1-phenylethoxy)-styrene and 0.080g AIBN in 16g toluene was heated in an oil bath at 75°C under inert atmosphere. After 48 hours the thick polymer solution was diluted with a minimum amount of toluene to allow its precipitation in 3L of petroleum ether. The recovered polymer (7.1g, 88% yield) had spectral properties in agreement with the proposed structure. Elemental analysis calculated for $(C_{16}H_{16}O)_n$: C, 85.67; H, 7.19; O, 7.13. Found: C, 85.80; H, 7.16. GPC. Mn: 17,000; Mw: 31,400

Imaging experiments. In typical experiments, the polymer solutions were prepared using the following proportions: 10g of polymer with 1.0g of triphenylphosphonium hexafluoroantimonate are dissolved into 53g of cyclohexanone at room temperature. The solution was spin coated onto 5" silicon wafers at 3200-3700 RPM adjusting the spinning speed to obtain films of 1μm thickness. The polymer films were then baked at 105°C for 10 minutes prior to exposure to 254 nm UV radiation through a mask using a Perkin-Elmer 500 deep-UV projection printing tool. After exposure the wafer was baked at 105°C to 140°C for 0.5-3 minutes and an image of the mask became visible. With polymer V, development of exposed areas was accomplished using alcoholic base and development of unexposed areas was accomplished using toluene-hexane mixtures.

Acknowledgments.

Financial support for this research was provided by the Natural Sciences and Engineering Research Council of Canada and by IBM Corporation; this support is acknowledged with thanks.

Literature Cited.

1. T.G. Tessier, J.M.J. Fréchet, C.G. Willson and H. Ito. Chapter 13 in "Materials for Microlithography" (L.F. Thompson, C.G. Willson and J.M.J. Fréchet, Editors) ACS Symposium Series, Washington D.C. 1984, 269. J.E. Guillet, S.K.L. Li and H.C. Ng, Ibid., Chapter 6, 1984, 165. S.K.L. Li and J.E. Guillet, Macromolecules 1977, 10, 840.
2. J.M.J. Fréchet, T.G. Tessier, C.G. Willson and H. Ito, Macromolecules 1985, 18, 317.
3. H. Ito, C.G. Willson and J.M.J. Fréchet, U.S. Patent 4,491,628, 1985.
4. J.M.J. Fréchet, E. Eichler, C.G. Willson and H. Ito, Polymer 1983, 24, 995.
5. H. Ito, C.G. Willson, J.M.J. Fréchet, M.J. Farrall and E. Eichler, Macromolecules, 1983, 16, 510.
6. H. Ito, C.G. Willson and J.M.J. Fréchet, Proc. SPIE, 1987, 771, 24.
7. C.G. Willson, H. Ito, J.M.J. Fréchet, T.G. Tessier and F.M. Houlihan, J. Electrochem. Soc., 1986, 133, 181.

8.   F.L. Benton and T.E. Dillon, J. Am. Chem. Soc., 1942, 64, 1128.
9.   H.C. Beyerman, and J.S. Bontekoe, Recl. Trav. Chim. Pays-Bas, 1962, 81, 691.
10. F. Houlihan, F. Bouchard, J.M.J. Fréchet, and C.G. Willson, Can. J. Chem., 1985, 63, 153.
11. J.M.J. Fréchet, B. Kryczka, F.M. Houlihan and E. Eichler, Manuscript in preparation. F.M. Houlihan, Ph.D. Dissertation, University of Ottawa, Ottawa, Canada, 1985.

RECEIVED July 13, 1988

Chapter 11

X-ray-Sensitive Alternating Copolymers

S. R. Turner[1], C. C. Anderson[1], K. M. Kolterman[1], and D. Seligson[2]

[1]Corporate Research Laboratories, Eastman Kodak Company, Rochester, NY 14650
[2]Intel Corporation, Santa Clara, CA 95051

A variety of alternating copolymers based on N-allyl- and N-(3-ethynylphenyl)maleimides, with substituted styrenes and vinyl ethers, have been prepared and their response to x-ray irradiation studied. Broadband and monochromatic x-ray exposures were conducted at the Stanford Synchrotron Radiation Laboratory. Sensitivities were observed to correlate with mass absorption coefficients of the copolymers and were found to be as high as 5-10 mJ/cm^2. Preliminary fine line lithographic studies indicate 0.5 μm resolution capabilities.

X-ray lithography (XRL) was first proposed by Spears and Smith in 1972 as a technique suitable for the high-volume manufacture of sub-micron-scale devices (<u>1</u>). It is only recently, however, that XRL has been considered as the most promising successor to optical lithography. Although several key issues such as the development of a suitable mask technology and the availability of an appropriate x-ray source are still being addressed, it has been predicted that XRL will see reasonable-volume, pilot-premanufacturing testing for 0.5 μm devices as early as 1988 (<u>2</u>). To meet this timetable, it is essential that new resist materials be developed which will be compatible with present and future x-ray sources. Trends in x-ray resists have recently been reviewed (<u>3</u>). In this report, we describe the preparation of a variety of copolymers based on N-allyl- and N-(3-ethynylphenyl)maleimides, with substituted styrenes and vinyl ethers and their evaluation as possible resists for x-ray lithography.

<u>Background</u>

Because the available photon flux and photon energy distribution for the various x-ray sources vary widely, the type of x-ray source utilized in an XRL exposure tool has a significant impact on resist selection. For this reason, it is appropriate to begin a report on resist design with a short summary describing the types of x-ray sources which are currently, or soon to be, available. A comparison

0097–6156/89/0381–0172$06.00/0
© 1989 American Chemical Society

of the important characteristics of the various sources is given in
Table I (4-6).

Table I.   Comparison of X-ray Sources

	Electron Impact	Plasma	Synchrotron
Flux (mW/cm^2)	0.1-0.2	1-5	> 50
Resolution (μm)	< 0.25	< 0.25	< 0.25
Photon energy	Pd source: 2.8	Ne plasma: 0.9-1.5	broadband
(keV)	W source: 1.7	Kr plasma: 1.6-2.0	400 eV-3 keV
Estimated cost	$1M	$1.2M	$1M per port
			8-10 ports/ring

X-ray sources have traditionally used electron-beam bombardment of
stationary or rotating metal (anode) targets to generate characteristic radiation and higher energy bremstrahlung. Commonly used
materials for x-ray targets include aluminum, copper, palladium,
silicon, and tungsten. Due to the low x-ray conversion efficiencies of these sources they provide a photon flux of only 0.10-0.20
mW/cm^2, which necessitates the use of extremely fast resists. To
generate the substantially higher intensities necessary for high
throughput XRL, two types of bright x-ray sources have been developed, plasma and synchrotron sources. Plasma sources rely on the
generation of a dense, hot plasma, either by magnetic compression
of ionized gases (z-pinch or plasma focus sources) or high-intensity
laser irradiation of solid surfaces (laser plasma source). Such
sources emit spectral radiation with photon energies of about 1-3
keV and intensities of 1-5 mW/cm^2 or more. The intense, broadband
radiation of synchrotron sources enable them to be used even with
slow resists such as standard optical resists or poly(methyl
methacrylate) (4).
     Plasma sources are expected to see production use in the early
1990's and synchrotron sources are not expected to make an impact
on commercial device fabrication in the United States until the mid
1990's (5). It appears that the first use of XRL for high-volume
device fabrication will rely on electron impact sources.

Discussion

Resist Design and Preparation. Our objectives were to develop new
resist materials which would be compatible with first generation
(electron impact) and second generation (plasma) x-ray sources.
The initial design goals for resist performance were chosen to
satisfy predicted production needs for XRL (6, 7); several of the
most critical criteria are summarized below.

sensitivity:	5-10 mJ/cm^2
resolution:	0.5 μm
contrast:	1.5
image stability:	> 150°C
etch resistance:	2X that for PMMA
tone:	positive or negative

The extremely high sensitivity requirements (as a point of reference
for resist sensitivity, consider that the sensitivity for poly-
(methyl methacrylate) positive resist and poly(chloromethylstyrene)
negative resist are 1500 and 60 mJ/cm^2, respectively) for the resist
system dictated the use of a negative working (cross-linking)
resist. The sensitivity of a negative resist depends on its radiation chemical yield for crosslinking, G_x, initial molecular weight,
M_w, and absorption of energy (8).

$$\text{Incident dose for gelation, } D_g = \frac{k \ N_A}{\rho t} \ \frac{1}{M_w \ G_x \ \mu_{mp}} \tag{1}$$

where N_A is Avogadro's number, k is a constant which depends on the
initial molecular weight distribution (k = 50 for a most probable
molecular weight distribution), ρ is the resist density, t is the
resist thickness, and μ_{mp} is the resist mass absorption coefficient.
For a polymer resist, consisting of i elements of mass absorption
coefficient μ_i and atomic weight A_i, the mass absorption coefficient
is given by:

$$\mu_{mp} = \frac{\sum_i A_i \mu_i}{\sum_i A_i} \tag{2}$$

     As can be seen from Equations 1 and 2, and as one would expect
intuitively, sensitivity is enhanced by the presence of highly
reactive groups (large G_x), high initial molecular weight, and the
incorporation of highly absorbing elements such as halogens or
metals. Other generalizations concerning the effect of polymer
structure on several resist performance parameters are summarized
in Table II. These guidelines are not entirely straightforward,
however, since several of the synthetic approaches for achieving
high sensitivity conflict with those for obtaining the best etch
resistance or resolution. Clearly, some compromises must be made
to obtain optimum overall performance. This may be accomplished by
the use of copolymers, blends, or multilayer schemes.

Table II. Effect of Structure on Resist Performance

Best Sensitivity	Best Etch Resistance	Best Resolution
Negative working (cross-linking) resist	Aromatic groups	Nonswelling resist —low initial M_w —optimize developer —high polymer T_g

High initial M_w
Highly reactive groups
Incorporate highly
  absorbing elements
  —halogens, metals
	High x-link density
	Multilayer schemes
	Incorporate
	  —I, Si for O plasma
	  —Sn for O and
	  F plasma
		Low sensitivity

No aromatic groups		

Previous work on allyl methacrylate homopolymers and statistical copolymers clearly showed that the allyl group is a very effective cross-linking function when exposed to x-rays (9). Resists with sensitivities of about 20 mJ/cm^2 were obtained. However, these polymers had only moderate contrasts ($\gamma \approx 1.1$) and poor dry etch resistance. In the work reported here, we have prepared and evaluated a variety of copolymers containing maleimide units with pendant allyl or ethynyl groups (structures Ia,b). The acronyms, structures, and molecular weights for the copolymers are tabulated in Table III. These copolymers, which are predominantly alternating copolymers, were attractive resist candidates for several reasons. The pendant allyl or ethynyl group provides a highly reactive cross-linking site. The substituent incorporated onto the styrene phenyl ring can be selected to obtain enhanced x-ray absorption at specific photon energies. This is illustrated in Figures 1 and 2 which show the predicted mass absorption coefficient (calculated from x-ray cross-section data using Equation 2) as a function of photon energy for a variety of N-allyl maleimide-substituted styrene copolymers. The substituent incorporated into the styrene can also provide improved resistance to oxygen (e.g., silyl-, stannyl-, or iodo-styrene) or fluorocarbon (e.g., stannyl-styrene) reactive-ion etching or provide an additional cross-linking site. These copolymers were also expected to be high T_g materials with the good chlorine dry etch resistance typical of cross-linked, aromatic polymers.

Ia   Ib

These copolymers are prepared by the solution free radical polymerization of the electron-poor monomer (substituted maleimide) and the electron-rich monomer (substituted styrene or vinyl ether). Predominantly alternating copolymers result from such polymerizations (10). We will report on this unique copolymerization that permits the copolymerization of two double bonds in the presence of a third reactive double bond elsewhere.

The alternating tendency of the copolymers is advantageous in that the polymerizations can be carried out to high conversions with little or no compositional drift. For random copolymerizations in which there is preferential incorporation of one monomer due to a mismatch in reactivity ratios, the compositional variations with conversion can be substantial. Such compositional heterogeneities in resist materials can lead to severe problems during image development.

Table III. Properties of Copolymers

Sample	Structure	$\overline{M}_n$	$\overline{M}_w$	$\overline{M}_w/\overline{M}_n$
MI-VBB-a	[structure: maleimide–N-allyl / styrene-CH$_2$Br copolymer]	12,730	246,380	19.35
MI-VBC-a	[structure: maleimide–N-allyl / styrene-CH$_2$Cl copolymer]	16,010	288,650	18.03
-b		35,480	– broad –	
-c		13,370	22,100	1.65
-d		22,330	86,190	3.86
-e		22,850	41,840	1.83
-f		33,590	92,210	2.75
-g		43,070	192,320	4.47
MI-VBI-a	[structure: maleimide–N-allyl / styrene-CH$_2$I copolymer]	40,630	40,630	16.58
MI-VBSi-a	[structure: maleimide–N-allyl / styrene-CH$_2$Si(CH$_3$)$_3$ copolymer]	21,270	220,140	10.35
MI-VBSn-a	[structure: maleimide–N-allyl / styrene-CH$_2$Sn(CH$_3$)$_3$ copolymer]	29,150	356,220	12.22
MI-4BrS-a	[structure: maleimide–N-allyl / 4-Br-styrene copolymer]	48,280	–	–
MI-BVE-a	[structure: maleimide–N-allyl / vinyl ether (CH$_2$)$_3$CH$_3$ copolymer]	13,170	65,770	4.99
-b		7,490	15,910	2.12
MI-CIEVE-a	[structure: maleimide–N-allyl / vinyl ether CH$_2$CH$_2$Cl copolymer]	17,050	–	–
-b		6,460	11,460	1.77
MI-DiBrS-a	[structure: maleimide–N-allyl / 3,5-diBr-styrene copolymer]	42,330	–	–
-b		18,930	60,460	3.19
MI-EVE-a	[structure: maleimide–N-allyl / vinyl ether CH$_2$CH$_3$ copolymer]	10,430	20,560	1.97

Table III. Continued. Properties of Copolymers

Sample	Structure	$\bar{M}_n$	$\bar{M}_w$	$\bar{M}_w/\bar{M}_n$
MI-PBrS-a		4,300	6,220	1.45
MI-VBASi-a		24,050	47,520	1.98
MI-S-a		24,400	55,520	2.28
EMI-VBC-a		31,140	81,570	2.62
PhAM-VBC-a -b		13,860 21,030	24,350 –	1.76 –
PhAMBr-VBC-a -b		16,230 28,220	– –	– –

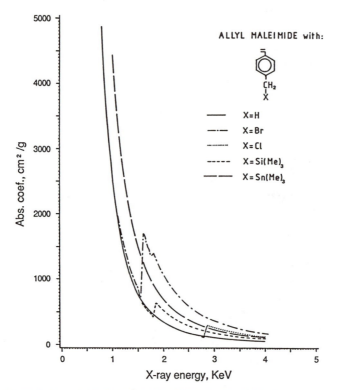

Figure 1.   Calculated mass absorption coefficient vs. x-ray energy for N-allylmaleimide copolymers.

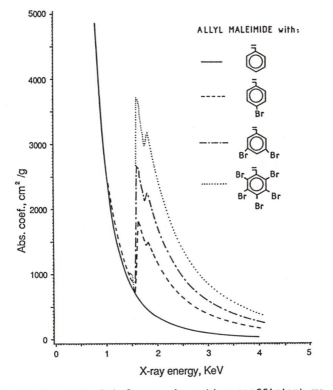

Figure 2. Calculated mass absorption coefficient vs. x-ray energy for N-allylmaleimide-brominated styrene copolymers.

Resist Evaluation. The x-ray exposures were performed at the lithography beam line at the Stanford Synchrotron x-ray source (SSRL). The storage ring was operated at 3 GeV with a beam current between 30 and 100 mA. Both white light (broadband) and monochromatic exposures were performed. The broadband spectrum has fairly constant power density between 0.8 and 3.5 keV with hard cutoffs above and below this range. A beryllium foil filter and a grazing-incidence gold mirror in the beam line are used to modify the synchrotron radiation spectrum from the ring as shown in Figure 3. Monochromatic exposures, using a layered synthetic microstructure monochromator (11), were made at several energies between 1.0 and 3.0 keV (the bandwidth for the monochromatic exposures is approximately $\pm$ 30 eV).

Contrast curves were obtained for each resist by measuring the thickness after development of a series of 1 mm by 5 mm exposed areas; the exposure dose typically varied from approximately 1 mJ/cm^2 to several J/cm^2 for the slowest resists. The majority of the resists were developed in ethyl acetate for 30 to 60 sec followed by a 20-sec rinse in 2-propanol. Initially, THF or a THF/2-propanol mixture was used as the developer; they were replaced by ethyl acetate because it provided superior contrast. Resist sensitivity was taken to be the incident dose which resulted in 50% exposed thickness remaining after development, $D_g^{.5}$. This is the standard convention for a negative resist.

Table IV summarizes the results for the resist screening studies at SSRL. The resist candidates generally exhibited good sensitivity, $D_g^{.5} < 50$ mJ/cm^2, and contrast, $\gamma = 1.5$. Definitive conclusions regarding the effect of structure on sensitivity are made difficult by the fact that polymer molecular weight and poly-dispersity varied considerably from sample to sample and, since we found a very strong dependence of $D_g^{.5}$ on these parameters, specific structural effects are obscured. Several generalizations can be made, however.

Table IV. Results for X-ray Exposures at SSRL

Sample	$D_g^{.5}$ mJ/cm	Contrast
P(MI-VBC)-b	15	1.5
P(MI-VBC)-a	2	1.3
P(EMI-VBC)-a	200	1.5
P(MI-VBB)-a	57	1.6
P(MI-4BrS)-a	5	1.0
P(MI-DiBrS)-a	10	1.7
P(MI-VBSn)-a	28	1.4
P(MI-VBSi)-a	85	1.1
PCMS (Japan Syn. Rubber Co.)	60	1.75
OEBR-100	100	1.4
polyglycidyl methacrylate (Tokyo Ohka Kogyo Co.)		
COP	20	1.5
poly(glycidyl methacrylate-co-ethyl acetate) (AT&T)		

As expected, the incorporation of pendant unsaturation in the resists greatly enhances sensitivity as demonstrated by a comparison of the contrast curves for poly(N-allyl maleimide-VBC) and the structurally similar poly(N-ethyl maleimide-VBC) (Figure 4). Both polymers have similar molecular weights and nearly identical mass absorption coefficients but the allyl-containing copolymer is 5X faster.

As previously mentioned, molecular weight has a significant impact on sensitivity. This can be seen from the $D_g^{.5}$ values for a series of poly(N-allyl maleimide-VBC) resists (obtained by fractionation of a high molecular weight, broad MWD sample). The sensitivity of these resists (i.e., $1/D_g^{.5}$) shows a linear dependence on M_n (see Figure 5). Resist samples which contained a substantial fraction of very high molecular weight polymer gave sensitivities much higher than one would predict from their M_n (see Figure 6). Although these resists provided exceptional speed, $D_g^{.5}$ = 2-10 mJ/cm^2, visual inspection of the wafer after development often showed scum remaining in the unexposed areas.

The effect of increased x-ray absorption on sensitivity was explored by conducting monochromatic exposures of a bromine-containing resist, poly(N-allyl maleimide-vinyl benzyl bromide), at photon energies which bracket the bromine absorption edges between 1.6 and 1.8 keV; contrast curves obtained for these monochromatic exposures are shown in Figure 7. The results are also plotted as $1/D_g^{.5}$ vs absorption coefficient in Figure 8; the data accurately follow the predicted inverse relationship defined by Equation 1.

The most promising resist materials identified in this study are copolymers of N-allyl maleimide with halogenated styrenes; of these, P(MI-DiBrS) provided the best combination of speed, $D_g^{.5}$ = 10 mJ/cm^2, and contrast, γ = 1.7. This resist gave superior performance compared to three commercially available negative e-beam/x-ray resists as demonstrated by the contrast curves in Figure 9. P(MI-DiBrS) gave a sensitivity 2-10 times higher than the commercial resists while providing equivalent or superior contrast. In addition, the high T_g for P(MI-DiBrS), 165°C, should make it more amenable to plasma processing than the commercial resists which have substantially lower T_g's (< 120°C). P(MI-DiBrS) also exhibits excellent x-ray absorption over a wide range of photon energies. This is illustrated in Figure 10 which compares the calculated mass absorption coefficient for this resist with the spectral output for a variety of x-ray sources. The resist has excellent absorption in the photon energy range characteristic of both Ne (0.9-1.5 keV) and Kr (1.6-2.0 keV) plasmas, storage rings (typically 0.4-4 keV), and it is especially well suited for the tungsten-target electron impact source utilized in the Perkin-Elmer stepper destined for the VHSIC x-ray lithography program. At the tungsten M_α line (1.67 keV), the absorption coefficient for P(MI-DiBrS) is about 2700 cm^2/g, which is more than 5X that for OEBR-100 or COP.

The dry etch resistance of P(MI-VBC) was compared with that for several commercial resists using a variety of etch gases (O_2, SF_6, and CF_4) under RIE conditions. The etch rate data given in Table V are presented as the ratio of the resist etch rate relative

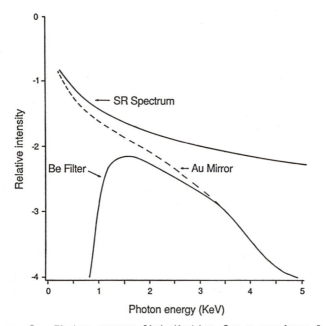

Figure 3.   Photon energy distribution for x-ray beam line at SSRL.

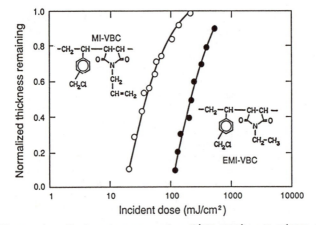

Figure 4.   Contrast curves for P(MI-VBC) and P(EMI-VBC).

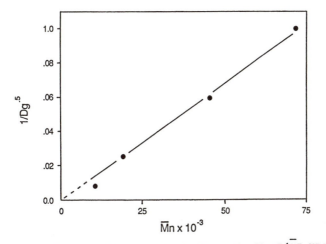

Figure 5. Sensitivity as a function of M_n for P($\overline{MI}$-VBC) resists.

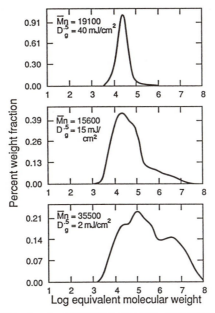

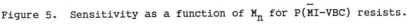

Figure 6. Effect of molecular weight distribution on sensitivity for P(MI-VBC) resists.

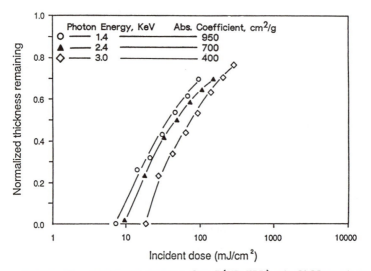

Figure 7.   Contrast curves for P(MI-VBB) at different x-ray energies.

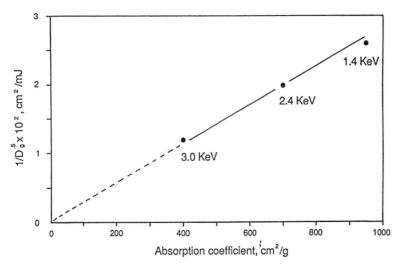

Figure 8.   Sensitivity as a function of mass absorption coefficient for P(MI-VBB).

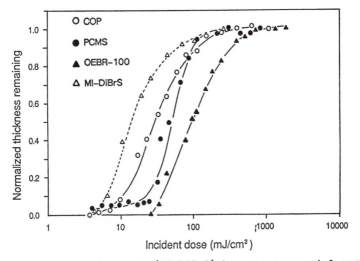

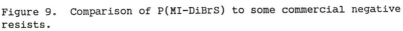

Figure 9. Comparison of P(MI-DiBrS) to some commercial negative resists.

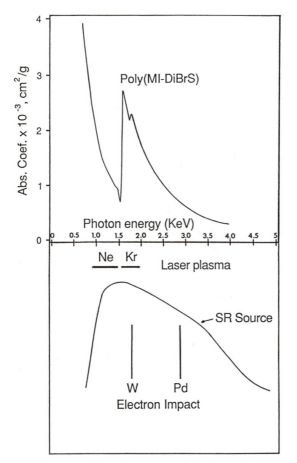

Figure 10. Comparison of the x-ray absorption of P(MI-DiBrS) to the spectral output of various x-ray sources.

to AZ1350 resist (O_2 RIE) or tungsten (SF_6 and CF_4 RIE). P(MI-VBC) exhibits the good dry-etch durability typical of aromatic polymers. Its etch rate under RIE conditions is comparable to that of the novolac-based photoresists AZ1350 and AZ5214. P(MI-VBC) which is an equimolar copolymer of allyl maleimide and chloromethylstyrene, has somewhat poorer etch durability compared with chloromethylstyrene homopolymer PCMS, but dramatically improved resistance to O_2, SF_6, and CF_4 RIE compared with the commercial negative x-ray resists OEBR-100 and COP. These results were obtained for unexposed resist films. Undoubtedly, better etch performance would be observed in exposed (cross-linked) films.

Table V. Dry Etch Performance

Resist	O_2 Resist Rate AZ1350 Rate	SF_6 Resist Rate Tungsten Rate	CF_4 Resist Rate Tungsten Rate
P(MI-VBC)	1.2	0.5	4.2
PCMS	1.0	0.45	3.3
OEBR-100	2.0	0.75	6.8
COP	1.8	0.73	7.0
PMMA	>2.0	0.75	5.6
AZ1350	1.0	0.5	4.0
AZ5214	0.75	0.55	4.2

Note: 10 mTorr, unexposed resists

No fine line exposures were performed at SSRL because a suitable x-ray mask was not available. However, the two best resist candidates based on the SSRL studies, P(MI-DiBrS) and P(MI-VBC), were imaged with a Pd-target x-ray stepper. The images shown in Figure 11 indicate that both resists are capable of at least 0.5 μm resolution.

Experimental

All materials used were analytical grade, obtained from Eastman Kodak Company, and used without further purification unless stated otherwise. Product purity was determined by NMR, GC, MS, elemental analysis, or comparison of melting points with literature values.

Monomer Preparation. Allyl Maleamic Acid--Allyl amine (84.4 g; 1.481 mol) in 100 mL of dichloromethane was added dropwise to recrystallized maleic anhydride (132 g; 1.346 mol) in 500 mL of dichloromethane at room temperature. The reaction mixture was stirred for 2 hr at room temperature and then filtered to obtain 206 g (99% of theory) of pale yellow crystals, m.p.: 110-112°C.

Allyl Maleimide--Allyl maleamic acid (50 g; 0.3326 mol) was melted at 125°C, poured into a 100 mL round bottom flask and distilled under reduced pressure to yield 17.8 g (39% of theory) of a colorless liquid which crystallized in the receiving flask as a white solid, m.p.: 41-44°C (Lit. m.p.: 42-44°C) (12).

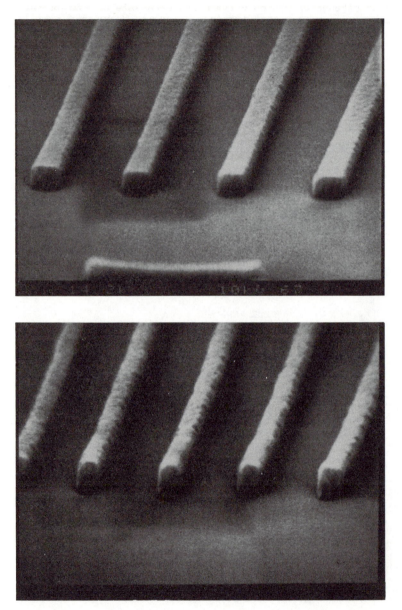

Figure 11.   Micrographs for P(MI-DiBrS). Images made with Pd(4.37 Å) source. Top: 0.6 µm; 180 mJ/cm^2 and bottom: 0.33 µm; 180 mJ/cm^2.

N-(3-ethynylphenyl)maleamic Acid--m-Aminophenylacetylene (25 g;
0.213 mol) in 50 mL of toluene was added dropwise to recrystallized
maleic anhydride (20 g; 0.204 mol) in 100 mL of toluene at room
temperature. The solution was stirred at room temperature for 2 hr
and then filtered to yield 41.3 g (94% of theory) of a fine gray
powder, m.p.: 194-196°C.

N-(3-ethynylphenyl)maleimide--N-(3-ethynylphenyl)maleamic acid
(21.5 g; 0.100 mol), anhydrous sodium acetate (6.0 g; 0.0731 mol),
and 175 mL of acetic anhydride (1.86 mol) were stirred together and
heated at 50°C for 3 hr. The reaction mixture was cooled to room
temperature and filtered. The filtrate was precipitated in an ice-
water mixture to yield an impure product. This product was dissolved in hot ethanol and recrystallized by addition of cold water
to yield 5.3 g (27% of theory) of a fine yellow powder, m.p.:
130-131°C.

3,5-Dibromostyrene--3,5-Dibromobenzyltriphenylphosphonium
bromide (125 g; 0.2054 mol) was added gradually to 830 mL (50-fold
excess) of 40% formaldehyde. 490 mL (30-fold excess) of 50% NaOH
was added dropwise. The 2000 mL round bottom flask was placed in a
cold water bath to limit the reaction exotherm. After addition of
the NaOH was completed, the reaction mixture was stirred at room
temperature for 2 hr. The mixture was filtered and washed with
ligroin to remove the triphenylphosphonium oxide by-product. After
drying over magnesium sulfate, the ligroin was removed by rotary
evaporation. The resulting liquid was placed in 150 mL of chilled
ethanol and 3,5-dibromostyrene precipitated as long white crystals.
Yield: 29 g (54% of theory), m.p.: 27-28°C.

Trimethyl Vinylbenzyl Silane--All steps to insure anhydrous
conditions necessary for a Grignard reaction were taken. Magnesium
turnings (10 g; 0.4112 mol) and 1000 mL anhydrous diethyl ether
were placed in a 2000 mL 3-necked round bottom flask equipped with
a reflux condenser and a closed nitrogen system. A few iodine
crystals and several drops of ethyl iodide were added to initiate
the reaction. Several drops of vinylbenzyl chloride (10 g; 0.0655
mol) in 10 mL ether were added to the flask. The mixture was heated
to reflux temperature. As the characteristic iodine color began to
disappear, the remainder of the vinylbenzyl chloride was added dropwise and the mixture was refluxed for 3 hr. After completion of
the Grignard reaction, chlorotrimethylsilane (71.3 g; 0.6568 mol)
in 50 mL of ether was added dropwise. The reaction mixture was
refluxed for an additional 14 hr. The resulting slurry was
filtered, washed in succession with saturated ammonium chloride
solution, 0.5% aqueous NaOH, and water, and then dried over
magnesium sulfate. The ether was removed by rotary evaporation and
distillation of the residue under reduced pressure gave 8.0 g (64%
of theory) of a pale yellow liquid.

Trimethyl Vinylbenzyl Stannane--This procedure is identical to
the preparation of trimethyl vinylbenzyl silane with the substitution of trimethyl tin chloride for the chlorotrimethyl silane. The
following quantities were used: 32.6 g (0.2136 mol) vinylbenzyl
chloride in 65 mL of ether, 7.81 g (0.3211 mol) magnesium turnings,

800 mL anhydrous ether, and 60 g (0.3015 mol) trimethyl tin chloride in 50 mL of ether. Yield: 26.4 g (44% of theory) of a pale yellow liquid after two distillations under reduced pressure.

Vinylbenzyl Bromide--Vinylbenzyl chloride (30 g; 0.196 mol) and tetrabutylammonium bisulfate (2.2 g; 0.007 mol) were added to a solution of sodium bromide (210 g; 1.96 mol) in 500 mL of distilled water and stirred for 5.5 hr at 50°C. The mixture was cooled to room temperature and the aqueous phase was separated from the organic phase and washed with dichloromethane. The two organic phases were combined, dried over magnesium sulfate, stripped of solvent, and stirred at 50°C with additional tetrabutylammonium bisulfate (0.71 g; 0.002 mol) and sodium bromide (72.1 g; 0.700 mol) in 70 mL of distilled water for 16 hr. The isolation process was repeated and distillation of the organic phase under reduced pressure (92°C/1.5 mm Hg) gave 25.5 g (66% of theory) of a pale yellow liquid.

Vinylbenzyl Iodide--Vinylbenzyl chloride (20 g; 0.131 mol) was added dropwise to dry sodium iodide (29.5 g; 0.198 mol) in 130 mL dry acetone. The mixture was stirred at 50°C for 40 min, cooled to room temperature, and filtered. The acetone was removed by rotary evaporation, and 100 mL water and 150 mL ether were added to the solid residue. The aqueous layer was washed with ether. The combined ether layers were washed with water containing 2% sodium thiosulfate and dried over magnesium sulfate. The ether was removed by rotary evaporation and the yellow residue was dissolved in 50 mL hexane and cooled to -20°C. Within 1.5 hr, yellow crystals formed. Fast filtering with chilled glassware provided 17.1 g (53.5% of theory) of vinylbenzyl iodide.

Polymer Synthesis. General Procedure--All polymers were prepared by free-radical-initiated solution polymerization. Typical quantities utilized were as follows: 5.0 g total monomer and 0.02 g AIBN or Vazo 33 in 30-60 mL solvent. More dilute solutions were employed in some cases to eliminate gel formation. In addition, a chain transfer agent, dodecanethiol, was used to control molecular weight in some polymerizations.

The monomers and solvent were placed in a round-bottomed flask and purged with nitrogen for 10-15 min. The mixture was heated to reflux temperature under nitrogen atmosphere. The initiator was then added in one portion along with the chain transfer agent. Reflux under nitrogen was continued for 5-24 hr. The mixture was cooled and the polymer was precipitated in a non-solvent.

Conclusions

Alternating copolymers of N-allyl maleimide with a substituted styrene are promising negative working x-ray resists. These materials have sensitivities of 10-50 mJ/cm^2, providing a contrast of 1.5 or better, and appear capable of at least 0.5 μm resolution. Further improvements in resist performance should be possible by optimization of polymer M_w and development conditions.

Acknowledgments

The authors wish to thank J. Reiff and M. Thomas for analytical support and D. R. Thompson for synthetic support.

Literature Cited

1. Spears, D. L.; Smith, H. I., <u>Solid State Technol.</u> 1972, 15, 21.
2. Heuberger, A., <u>Microelectron. Eng.</u> 1985, 3, 535.
3. Taylor, G. N., <u>Solid State Technol.</u> 1984, June, 124.
4. Lyman, J., <u>Electronics</u>, March 17, 1986, 46.
5. Lyman, J., <u>Electronics</u>, December 2, 1985, 45.
6. Wilson, A. D., <u>Proc. SPIE, Int. Soc. Opt. Eng.</u>, 1985, 537, 85.
7. Glendinning, W. B., <u>Solid State Technol.</u> 1986, 97.
8. Schnabel, W., <u>Prog. Polym. Sci.</u>, 1983, 9, 297.
9. Daly, R. C.; Hanrahan, M. J.; Blevins, R. W., <u>Proc. SPIE. Int. Soc. Opt. Eng.</u> 1985, 539, 138.
10. Cowie, J. M. G. (editor), <u>Alternating Copolymers</u>, Plenum Press, 1985.
11. Pianetta, P.; Redaelli, R.; Jaeger, R.; Barbee, T. W., <u>Proc. SPIE, Int. Soc. Opt. Eng.</u>, 1985, 537, 69.
12. Pyriadi, T. M. and Harwood, H. J., <u>Polymer Preprints</u>, 1970, 11(1) 60.

RECEIVED August 4, 1988

Chapter 12

Effect of Proton Beam Energy on the Sensitivity and Contrast of Select Si-Containing Resists

Diana D. Granger, Leroy J. Miller, and Margaret M. Lewis

Hughes Research Laboratories, 3011 Malibu Canyon Road, Malibu, CA 90265

Poly[γ-methacryloyloxypropyltris(trimethylsiloxy)silane], or PMOTSS, is an intrinsically negative resist, although most poly(alkyl methacrylates) are positive resists. The inherent sensitivity of PMOTSS, expressed as the product of the electron beam gel dose and the weight-average molecular weight, $D_g^i \overline{M}_w$, is comparable to those of the polyhalostyrenes. Although the PMOTSS homopolymer is rubbery, copolymers of MOTSS and halostyrenes (HS) are hard and glassy and suitable for use as the top, imaging layer of a bilayer resist for electron beam and masked ion beam lithography. The sensitivity of each copolymer is a function of its molecular weight and the inherent sensitivity and mole fraction of the HS. For each resist, the half-thickness sensitivity decreased and the contrast increased with increasing proton beam energy over the range of 90 to 250 keV. The sensitivity decreased and the contrast increased further for irradiation with 20 keV electrons. The influence of the proton beam energy is most pronounced for those copolymers that have the poorest sensitivity.

Radiation-sensitive polymers are used to define pattern images for the fabrication of microelectronic devices and circuits. These polymers, called resists, respond to radiation by either chain scission (positive resists) or by crosslinking (negative resists). In positive resists, the exposed areas dissolve selectively by chemical developers; in negative resists, the exposed areas are insoluble and remain after development.

Although the single-layer negative resists commonly utilized are simple, they typically do not resolve narrow (0.5 μm) gaps between wide lines or pads due to proximity effects from backscatter from the substrate during electron beam (E-beam) exposure. Further, there are variations in linewidth which occur when images are written in a single-layer resist which overlies steps in the substrate.

0097–6156/89/0381–0192$06.00/0
© 1989 American Chemical Society

In order to achieve a higher resolution, a multilayer resist must be used. A preferred type is a bilevel resist that consists of a thick organic polymer bottom layer overcoated with a thin silicon-containing imaging layer. The bottom layer serves to planarize the substrate topography and to protect the top, imaging layer from exposure from backscattered electrons. After a pattern is written and developed in the top layer, the resulting image is used as a mask for transferring the pattern into the thick bottom layer by reactive ion etching with oxygen (O_2 RIE). This process is shown in the scheme in Figure 1. A final resist image can be obtained with vertical sidewalls and a high aspect ratio.

High resolution negative resists are needed for masked ion beam lithography (MIBL) and for the fabrication of MIBL masks by E-beam lithography (EBL). The MOTSS copolymer resists were developed to obtain the resolution of fine features that a bilevel resist can best provide. The flexibility afforded by choosing the structure of the HS, the copolymer composition, and the molecular weight allows a resist to be tailored by simple synthesis adjustments to have the particular sensitivity and etch protection which best suits the application.

The response of resist materials to ionizing radiation is an important factor in the lithographic process. Coincident with a sensitivity survey of various HS-MOTSS compositions, it was observed that an increase in proton beam energy resulted in a decrease in sensitivity and an increase in contrast *(1)*. Hall, Wagner, and Thompson reported an increase in sensitivity when positive or negative resists are exposed with a series of ions with increasing mass but identical energy *(2)*. The sensitivity increase is related to the increased energy dissipation along the particle track by the slower-moving, heavier ions *(3)*. In similar work by the same authors, it was noted that the contrast decreased for negative resists and increased for positive resists when exposed with the heavier particles *(4)*. This paper is an expansion of our earlier report *(1)* of the effect of the proton energy on the sensitivity and contrast of the HS-MOTSS copolymers.

Experimental

Materials.   The homopolymer PMOTSS and its copolymers with 4-chlorostyrene (4CS), 3-chlorostyrene (3CS), 3,4-dichlorostyrene (34DCS), 4-bromostyrene (4BS), 4-iodostyrene (4IS), and a 60:40 commercial mixture of 3- and 4-chloromethylstyrene (CMS) were prepared by free-radical polymerization in benzene solution at 65 °C to 80 °C with benzoyl peroxide (BPO) as the initiator. The total monomer concentration was 1.5M for all the copolymerizations, while the molar ratios of HS to MOTSS varied from 1 to 4. Initially, the monomers were used as received; however, when the control of the molecular weight became important, the inhibitor was removed from the monomers by passage through a column containing the De-hibit 100 ion exchange resin from Polysciences, Inc.

Elemental analyses were performed by Galbraith Laboratories, Inc., of Knoxville, TN. The compositions of the HS-MOTSS copolymers were derived from the average of the values calculated from the reported halogen and silicon content. The molecular weight and dispersity of the polymers were measured in tetrahydrofuran by size exclusion gel permeation chromatography against monodisperse polystyrene standards using a Waters Associates Model ALC-202/401 liquid chromatograph equipped with Ultrastyragel Model 10681 columns.

Exposure and Processing. E-beam sensitivity matrices were written using the AEBLE 150 E-beam system with 20 keV electrons at a flux of either 10 or 25 A/cm^2. The copolymers were tested as 0.5-μm spin-coated films on silicon wafers, which were typically prebaked at 60-80 °C at about 5 torr. The sensitivity to protons was measured using exposures in similar 0.5-μm films. A Hughes-built ion implantation system exposed a series of ten 1-mm diameter spots on each wafer covering a selected range of dosages.

The exposed HS-MOTSS copolymer films were dip-developed in solvent followed by baking at 50-80 °C in vacuum for about 30 min. E-beam dose matrice images were developed in chlorobenzene for 30 sec; ion beam test exposures were developed in methyl isobutyl ketone for 30 sec. The thicknesses of the films were measured with a Nanospec/AFT microarea film thickness measuring system.

Figure 2 shows a typical plot of normalized thickness remaining for a negative resist against the log of the radiation dose. The sensitivity, Q, for the resist is the dose which results in one-half of the original thickness remaining after development. The symbol Q_e refers to the E-beam sensitivity in $\mu C/cm^2$; Q_H refers to the proton sensitivity in ions/cm^2; and Q_p refers to the proton sensitivity in units of $\mu C/cm^2$ for comparison with Q_e. The contrast γ is the slope of the linear portion of the curve, which is typically within the normalized thickness range of 0.1 to 0.7. The gel dose D_g^i is the dose at which that line extrapolates to a normalized thickness of zero, and usually corresponds to the minimum dose for gelation.

Results and Discussion

The structure of the MOTSS monomer is:

$$
\begin{array}{c}
CH_3 \\
| \\
H_2C=C-C=O \\
|
\end{array}
\qquad
\begin{array}{c}
O-Si(CH_3)_3 \\
| \\
O-CH_2-CH_2-CH_2-Si-O-Si(CH_3)_3 \\
| \\
O-Si(CH_3)_3
\end{array}
$$

As a silicon-substituted poly(alkyl methacrylate), the homopolymer of MOTSS was expected to act as a positive resist (5). The fact that PMOTSS is a negative rather than a positive resist must be at least partly due to the large size of the ester alkyl group. Lai and Helbert have shown that the G_{Xlink} of poly(alkyl methacrylates) increases as the size of the ester group increases (6,7). Alternatively, the negative tone could arise from the chemistry of the trimethylsiloxy groups. It has been shown that trimethylsiloxy-terminated poly(dimethylsiloxanes) act as negative E-beam resists (8), as does poly(methyl methacrylate) with pendent poly(dimethylsiloxane) grafts (9).

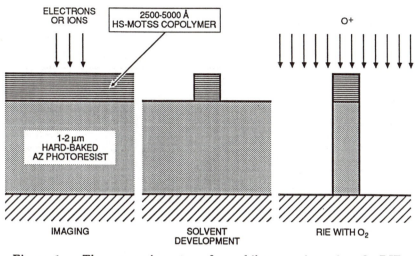

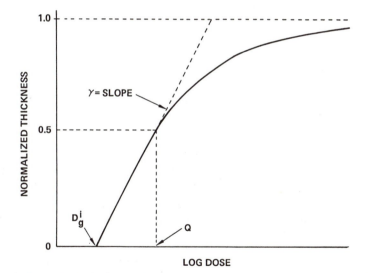

Figure 1. The processing steps for a bilayer resist using O_2-RIE.

Figure 2. A typical sensitivity curve for a negative resist showing the relationship between the gel dose (D_g^i), the sensitivity (Q), and the contrast (γ).

Table I.   The polymerization of MOTSS with various halostyrenes

Halostyrene (HS)	Temp °C	Time hr	[HS] M	[MOTSS] M	[BPO] Mx10³	% Conv.	HS/MOTSS Feed	Polymer
(PMOTSS)	80	6	0.00	0.311	2.6	88.9	0.00	0.00#
4CS	80	6	0.758	0.712	7.46*	53.7	1.06	1.26
	80	6	0.786	0.761	12.7	65.1	1.03	1.32#
	80	6	0.811	0.717	14.6*	69.6	1.13	1.52
	80	6	1.27	0.361	14.5	58.3	3.51	3.92
3CS	75	12	1.16	0.316	2.87	23.3	3.68	7.47#
34DCS	70	12	0.748	0.753	5.01	17.6	0.99	1.94#
4BS	80	6	1.17	0.335	5.15	29.6	3.50	4.27#
4IS	80	6	0.176	0.191	0.28	30.0	0.92	1.30
CMS	80	6	0.751	0.749	12.9	86.5	1.00	1.21
	80	6	0.811	0.714	12.9	86.1	1.14	1.32#
	80	12	0.779	0.751	7.52*	77.2	1.04	1.46
	80	12	0.784	0.714	3.78*	62.2	1.10	1.72
	80	6	0.752	0.728	12.9	>61.1	1.03	1.73
	80	12	0.885	0.617	10.2	74.5	1.43	2.08
	75	23	0.992	0.509	5.20	57.9	1.95	3.28
	80	6	1.21	0.325	6.69	75.5	3.72	4.57#
	64	260	1.21	0.300	1.7	22.1	4.02	8.65#

* Inhibitor removed from monomers.
# Material used for reported proton beam data.

Because it is rubbery at room temperature, the PMOTSS homopolymer is unsuitable for resist processing. To obtain radiation-sensitive materials with the etch protection afforded by the silicon content of MOTSS but with a higher softening point temperature, MOTSS was copolymerized with various halostyrenes. Table I summarizes the polymerization conditions for the homopolymer and copolymers, the molecular weights of which are listed in Table II. When polymerized at 80 °C, the product copolymers contained a slightly higher HS:MOTSS ratio than was provided in the reaction mixture, and this ratio increased when the temperature was decreased. The HS-MOTSS copolymers are hard and glassy and suitable for use as the top layer of bilevel resists. The softening point temperatures for a series of 4CS- and CMS-MOTSS copolymers were in the range of 44 to 78 °C and increased with increasing wt% HS content (1).

Table II. The molecular weights of the MOTSS copolymers

Halostyrene (HS)	$\frac{HS}{MOTSS}$	$\overline{M}_w \times 10^{-3}$	$\overline{M}_n \times 10^{-3}$	$\overline{M}_w/\overline{M}_n$
(PMOTSS)	0.00	76.8	41.9	1.83
4CS	1.26	36.8	25.6	1.44
	1.32	29.3	19.2	1.52
	1.52	30.1	20.8	1.44
	3.92	40.6	23.0	1.76
3CS	7.47	68.2	39.1	1.74
34DCS	1.94	140.	83.4	1.68
4BS	4.27	53.6	25.0	2.14
4IS	1.30	25.9	17.7	1.46
CMS	1.21	63.0	43.6	1.44
	1.32	112.	37.0	3.04
	1.46	50.2	32.0	1.57
	1.72	69.4	43.2	1.61
	1.73	58.5	41.1	1.42
	2.08	112.	57.6	1.94
	3.28	106.	36.4	2.91
	4.57	82.9	32.4	2.56
	8.65	190.	42.7	4.44

The E-beam sensitivity of an HS-MOTSS copolymer resist is controlled by the molecular weight of the copolymer and by the structure and weight fraction X of the HS. The product of the E-beam gel dose and the weight-average molecular weight, $D_g^i\overline{M}_w$, can be used as a figure of merit for negative resists; the lower the $D_g^i\overline{M}_w$, the more inherently sensitive the resist. The value of 0.31 reported *(1)* for the PMOTSS homopolymer is comparable to those of the halostyrene polymers *(10,11)*. The $D_g^i\overline{M}_w$ value for a copolymer can be predicted by the formula derived by Tanigaki et al. *(11)*:

$$\frac{1}{(D_g^i\overline{M}_w)_{HS-MOTSS}} = \frac{X}{(D_g^i\overline{M}_w)_{HS}} + \frac{1-X}{(D_g^i\overline{M}_w)_{MOTSS}}$$

The $D_g^i\overline{M}_w$ values for a series of CMS-MOTSS copolymer compositions are shown in Figure 3 with the lines calculated from the above formula using two $D_g^i\overline{M}_w$ values for the PCMS homopolymer. Extrapolation of the experimental data suggests that the actual $D_g^i\overline{M}_w$ value for the PCMS homopolymer is 0.057, midway between our earlier value of 0.046 (Jensen, J.E.; Brault, R.G.; Miller, L.J.; Granger, D.D.; van Ast, C.I., Hughes Research Laboratories, unpublished results) and that of 0.07 used by Tanigaki et al. *(11)*.

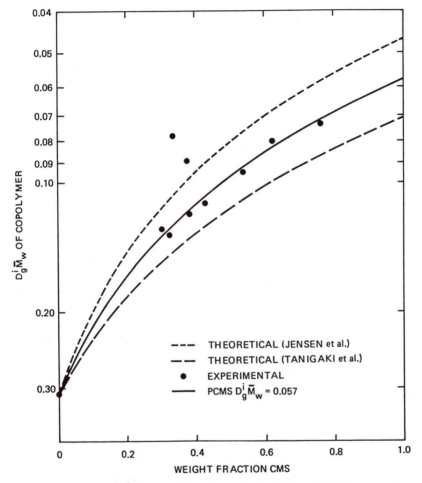

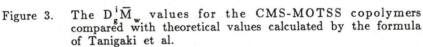

Figure 3.   The $D_g^i\bar{M}_w$ values for the CMS-MOTSS copolymers compared with theoretical values calculated by the formula of Tanigaki et al.

Table III. The inherent E-beam sensitivity of PMOTSS
and its copolymers

Halostyrene (HS)	$\dfrac{HS}{MOTSS}$	wt% HS	Predicted $D_g^i\bar{M}_w$ [a]	$D_g^i\bar{M}_w$ [b]	Experimental $D_g^i\bar{M}_w$
4CS	1.26	28.3	0.35	0.31	0.167
	1.32	30.1	0.35	0.31	0.214
	1.52	32.4	0.36	0.31	0.252
	3.92	56.3	0.40	0.32	0.644
3CS	7.47	70.5	0.28	-	0.400
34DCS	1.94	44.3	0.35	-	0.182
4BS	4.27	65.6	0.35	-	0.790
4IS	1.32	41.2	0.31	0.25	0.992
CMS	1.21	30.2	0.11	0.15	0.128
	1.32	32.2	0.11	0.15	0.132
	1.46	33.5	0.11	0.14	0.078
	1.72	37.5	0.098	0.14	0.089
	1.73	38.2	0.097	0.13	0.118
	2.08	42.6	0.090	0.12	0.111
	3.28	54.2	0.075	0.11	0.093
	4.57	62.2	0.068	0.099	0.080
	8.65	75.7	0.058	0.086	0.073

[a] $D_g^i\bar{M}_w$ values for P4CS, P3CS, P34DCS, P4BS, P4IS, and PCMS are 0.52, 0.27, 0.42, 0.38, 0.32, and 0.046, respectively (Jensen et al.).
[b] $D_g^i\bar{M}_w$ values for P4CS, P4IS, and PCMS are 0.32, 0.2, and 0.07, respectively *(11)*.

The predicted $D_g^i\bar{M}_w$ values for the HS-MOTSS copolymers are listed in Table III along with those obtained experimentally. While the actual $D_g^i\bar{M}_w$ values for the CMS-MOTSS copolymers followed the Tanigaki formula reasonably well, those for the other HS-MOTSS copolymers did not. It was anticipated *(11)* that those copolymers which had low HS content would be more sensitive than calculated; however, some of the copolymers which contained a larger weight fraction of HS were much less sensitive than predicted by the formula. (The departure from the Tanigaki formula in these cases could be an effect of scission processes in the MOTSS units. A reviewer mentioned that the determination of the values for $G_{scission}$ and G_{Xlink} for PMOTSS might be helpful in distinguishing between the degradation of the methacrylate chain and the crosslinking caused by the sidechains. Sensitivity curves for PMOTSS indicate that the exposed areas may begin to thin slightly as the E-beam dose is increased beyond about 40 $\mu C/cm^2$. Whether the thickness loss is due to chain scission or to the loss of sidechain groups is unclear; however, if the dose at which the material crosslinks approaches that at which it loses thickness, the net loss in sensitivity might produce a larger $D_g^i\bar{M}_w$ value than that predicted by the formula.)

The sensitivities to 90, 125, 175, and 250 keV protons for PMOTSS and a representative selection of its copolymers are listed in Table IV. In general the sensitivity of the materials decreased as the energy of the protons increased. This is true because the energy deposited in the film per unit of pathlength actually decreases as the energy of the incident particles is increased. Because a proton with higher kinetic energy moves faster through a resist film, spending less time near a given molecule in its path, there is a lower probability that it will transfer energy to that molecule than would a slower-moving, lower energy particle *(12)*.

Table IV.   The proton beam sensitivity of PMOTSS
and its copolymers

Halostyrene (HS)	$\dfrac{HS}{MOTSS}$	$Q_H \times 10^{-12}$ ions/cm^2				log-log slope	corre. coeff.
		90keV	125keV	175keV	250keV		
(PMOTSS)	0.00	1.37	1.55	1.63	1.89	0.299	0.988
4CS	1.32	2.38	2.68	3.01	3.43	0.372	0.999
3CS	7.47	1.54	2.00	2.01	2.09	0.268	0.841
34DCS	1.94	0.538	0.607	0.708	0.710	0.288	0.948
4BS	4.27	4.42	5.78	6.31	6.65	0.373	0.932
CMS	1.32	0.521	0.525	0.631	0.627	0.217	0.895
	4.57	0.636	0.616	0.695	0.757	0.190	0.900
	8.65	0.264	0.283	0.335	0.332	0.250	0.928

Q_H = Half-thickness proton beam sensitivity.

The region of exposure for an individual proton can be visualized as a cylinder which contains the deposited energy and which extends through the resist film with the path of the incident ion at the axis of the cylinder. As the proton moves through the polymer, it ionizes the polymer and generates secondary electrons which distribute the absorbed energy around the primary particle track. Of course, the density of deposited energy is higher at the center of the cylinder than it is near the circumference. The secondary electrons have average kinetic energies proportional to that of the incident proton; therefore, those generated by higher energy protons have longer ranges *(4)*, and they define cylinders with larger radii. There is a much lower density of deposited energy in the exposure cylinder for a higher energy proton because not only does its exposure cylinder contain less energy per unit depth than does that of a lower energy proton, but the longer ranges of the secondary electrons spread that energy throughout a larger volume. Conversely, the greater energy dissipation per unit pathlength of lower energy protons combines with the shorter range of the secondary electrons to result in a higher density of deposited energy within the exposure cylinder. The influence of the proton beam energy on the sensitivity is most pronounced for those copolymers that have the poorest sensitivity. Figure 4 presents least-squares fits of log-log plots of the sensitivity, Q_H, as a function of the proton energy. The slopes of these plots, listed in Table IV with their correlation coefficients, can be taken as a measure of the response, R_p, of the MOTSS copolymer to changes in proton energy and to changes in the amount of energy deposited within the resist by each proton. Figure 5 is a plot of the values of R_p as a function of the sensitivities of the corresponding polymers to 125 keV protons, and it demonstrates that R_p is largest for the least sensitive polymers. We believe that the diminished response of the more sensitive copolymers is related to the crosslinking efficiency of the deposited energy. If a resist has a low sensitivity, it can use the extra energy deposited by low energy protons effectively, even at a high energy density; but if the resist has a high sensitivity, the extra absorbed energy has a smaller effect because it is deposited at too high a density and most of it is wasted *(2,4,13)*.

Table V.   The proton beam contrasts for PMOTSS
and its copolymers

Halostyrene (HS)	$\dfrac{HS}{MOTSS}$	Contrast (γ_H)			
		90keV	125keV	175keV	250keV
(PMOTSS)	0.00	1.41	1.41	1.46	1.67
4CS	1.32	1.65	1.35	1.47	1.82
3CS	7.47	1.41	1.36	1.39	1.53
34DCS	1.94	0.980	1.06	1.06	1.16
4BS	4.27	1.69	1.75	1.49	2.50
CMS	1.32	0.676	0.714	0.810	0.800
	4.57	0.625	0.714	0.725	0.845
	8.65	0.645	0.613	0.645	0.674

The proton beam contrasts for the MOTSS copolymer resists are listed in Table V. The contrasts generally increased with increasing proton energy, although there is considerable scatter in the data, which reflects the difficulty in obtaining accurate values for the contrast. The increase in contrast (Figure 6) is believed to be related to the concomitant decrease in the density of the deposited energy within the exposure cylinders. Contrast is a measure of the efficiency of the incident radiation in effecting crosslinking after the gel point dose has been reached. If the energy is deposited at a density that is greater than is needed to crosslink the resist, some of the energy is wasted. As the proton energy is increased, the volume of the exposure cylinders increases and the average energy density decreases,

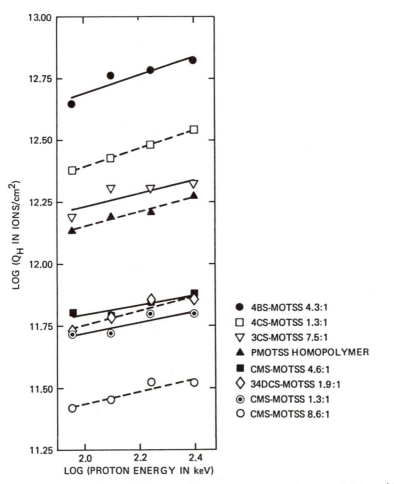

Figure 4.   The effect of proton beam energy on the sensitivity (Q_H) of the MOTSS copolymer resists.

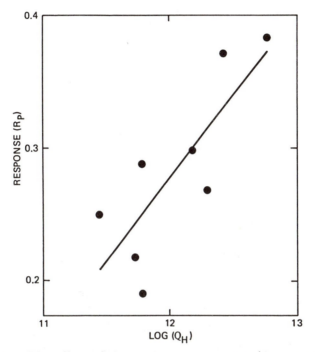

Figure 5.   The effect of proton beam sensitivity $(Q_H$ at 125 keV) on the response (R_p) of the MOTSS copolymer resists to changes in proton beam energy.

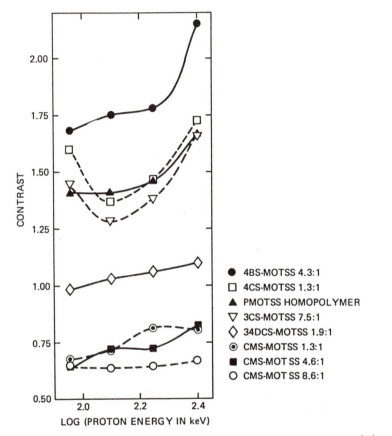

Figure 6.   The effect of proton beam energy on the contrast (γ) of the MOTSS copolymer resists.

resulting in a more efficient energy distribution for crosslinking the resist.

Table VI. The ratio of E-beam and proton beam sensitivities of PMOTSS and its copolymers

Halostyrene (HS)	$\dfrac{HS}{MOTSS}$	Q_e ($\mu C/cm^2$)	Q_e/Q_p 90keV	125keV	175keV	250keV
(PMOTSS)	0.00	7.58	34.6	30.6	29.1	25.1
4CS	1.32	11.6	30.5	27.0	24.1	21.1
3CS	7.47	10.3	41.8	32.2	32.0	30.8
34DCS	1.94	2.04	23.7	21.0	18.0	18.0
4BS	4.27	21.6	30.5	23.4	21.4	20.3
CMS	1.32	1.78	21.4	21.2	17.6	17.7
	4.57	2.13	20.9	21.6	19.2	17.6
	8.65	1.30	30.8	28.7	24.2	24.5
	Averages:		29.3	25.7	23.2	21.9

Q_e = E-beam sensitivity *(1)*; Q_H = proton beam sensitivity.
$Q_p = Q_H \times (9.65 \times 10^{10}\ \mu C/mole)\ /\ (6.02 \times 10^{23}\ ions/mole)$.

The copolymers' sensitivities to protons, expressed in $\mu C/cm^2$ (Q_p), are compared to their sensitivities to 20 keV electrons (Q_e) in Table VI. On the average, they were 25 times more sensitive to protons than to electrons, and Q_e/Q_p decreased as the proton energy increased. The contrasts, which are compared in Table VII, were on the average about 2 times as high for electrons as with protons, and the ratio γ_e/γ_p also decreased as the proton energy increased. The explanation for these results is consistent with that given above for the effects of proton energy. Only about 5% of the energy of the electron beam is dissipated in the resist *(4)*. The secondary electrons have energies approaching those of the primary electrons, and therefore they have much longer ranges than those generated by protons. Consequently, the electron beam energy absorbed by the resist is spread over a relatively large volume and the absorbed energy density is low, compared to protons.

An earlier paper from this laboratory showed that a plot of $\log(Q_e)$ at 20 keV against $\log(Q_p)$ at 100 keV for 19 negative and positive resists had a linear least squares slope of 1.30 with a correlation coefficient of 0.909 *(14)*. Our log-log plot of Q_e against Q_p at 90 and 125 keV for the MOTSS copolymers is shown in Figure 7. The linear least squares fit for the 125 keV plot has a slope of 1.05 with a correlation coefficient 0.992. Those for the 175 and 250 keV protons have the same slope within experimental error, but that for the 90 keV protons is higher, 1.11, with a correlation coefficient of 0.981. The reason that the slopes derived from the HS-MOTSS data

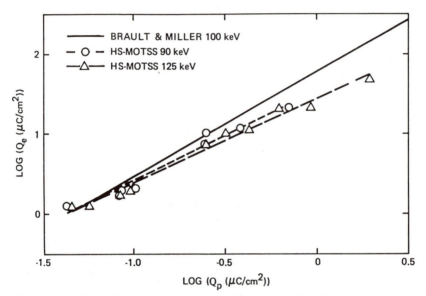

Figure 7.   The E-beam sensitivity (Q_e) of MOTSS copolymer resists as a function of the proton beam sensitivity (Q_p) with 90 and 125 keV protons.

are lower than that reported earlier for other resists is unclear and may be an effect of the $G_{scission}$ of the methacrylic MOTSS units; however, as noted in the previous paper *(14)*, a slope >1.00 reflects a greater crosslinking efficiency in the less sensitive resists.

Table VII.   The ratio of E-beam and proton beam contrasts for PMOTSS and its copolymers

Halostyrene (HS)	$\dfrac{HS}{MOTSS}$	γ_e	γ_e/γ_H			
			90keV	125keV	175keV	250keV
(PMOTSS)	0.00	1.79	1.27	1.27	1.23	1.07
4CS	1.32	2.27	1.42	1.66	1.54	1.31
3CS	7.47	2.15	1.49	1.68	1.56	1.30
34DCS	1.94	2.57	2.62	2.50	2.42	2.34
4BS	4.27	2.81	1.67	1.60	1.58	1.31
CMS	1.32	2.89	4.29	4.03	3.55	3.59
	4.57	1.81	2.85	2.51	2.50	2.19
	8.65	1.34	2.09	2.10	2.08	2.00
Averages:			2.21	2.17	2.06	1.89

γ_e = E-beam contrast; γ_H = Proton beam contrast.

In summary, we have shown that the proton beam sensitivity of an HS-MOTSS copolymer decreases and the contrast increases with increasing proton energy, and that the influence of the proton energy on the sensitivity of these materials is diminished with the more sensitive resists.

Acknowledgments

The authors wish to thank David H. Adair and Doug M. Jamba for providing the proton beam exposures, and Ken A. Barnett, Jane T. Castillo, Terry J. Huff, and Kathleen D. Miller, who provided the E-beam exposures under the direction of Oberdan W. Otto.

Literature Cited

1. Granger, D.D.; Miller, L.J.; Lewis, M.M. *J. Vac. Sci. Technol. B* **1988**, *6*, 370.
2. Hall, T.M.; Wagner, A.; Thompson, L.F. *J. Vac. Sci. Technol.* **1979**, *16*, 1889.
3. Komuro, M.; Atoda, N.; Kawakatsu, H. *J. Electrochem. Soc.* **1979**, *126*, 483.
4. Hall, T.M.; Wagner, A.; Thompson, L.F. *J. Appl. Phys.* **1982**, *53*, 3997.
5. Reichmanis, E.; Smolinsky, G. *Proc. SPIE* **1984**, *469*, 38.

6. Lai, J.H.; Helbert, J.H. *Macromolecules* **1978**, *11*, 617
7. Lai, J.H. *J. Imaging Technol.* **1985**, *11*, 164.
8. Babich, E.; Paraszczak, J.; Hatzakis, M.; Shaw, J. *Microelectronic Eng.* **1985**, *3*, 279.
9. Bowden, M.J.; Gozdz, A.S.; Klausner, C.; McGrath, J.E.; Smith, S. In *Polymers for High Technology: Electronics and Photonics*; Bowden, M.J.; Turner, S.R., Eds.; ACS Symposium Series No. 346; American Chemical Society: Washington, DC, 1987; pp 122-137.
10. Stinson, S.C. *Chem. Eng. News* **1983**, *61(32)*, 7.
11. Tanigaki, K.; Ohnishi, Y.; Fujiwara, S. *Prepr. Pap. - Am. Chem. Soc., Div. Ctgs. & Plstcs.* **1983**, *48*, 179.
12. Chapiro, A. *Radiation Chemistry of Polymeric Systems*; Wiley Interscience: New York, 1962; p 18.
13. Brown, W.L.; Venkatesan, T.; Wagner, A. *Nucl. Instrum. Methods* **1981**, *191*, 157.
14. Brault, R.G.; Miller, L.J. *Polym. Eng. Sci.* **1980**, *20*, 1064.

RECEIVED July 15, 1988

Chapter 13

Novel Principle of Image Recording

Photochemically Triggered Physical Amplification of Photoresponsiveness in Molecular Aggregate Systems

Shigeo Tazuke and Tomiki Ikeda

Research Laboratory of Resources Utilization, Tokyo Institute of Technology, 4259 Nagatsuta, Midori-ku, Yokohama 227, Japan

A novel principle of amplified photochemical reversible imaging is presented. This principle is based on the concept of photochemically triggered physical amplification of photoresponsiveness of molecular aggregate systems by means of inducing various phase transitions brought about by a partial photochemical change. Applicability of this principle was demonstrated for micelles, vesicles, liquid crystals, and finally for liquid crystalline polymer films. Owing to the presence of a threshold photoenergy to induce thermodynamic transition, the degree of image amplification relative to direct read-out by photochromism exceeded 10^2. Five examples are described.

Studies on image recording systems and their supporting materials which enable high density, high speed, highly reliable reversible information storage and read-out are topics of current material research. Since there seems to be a foreseeable limit of performance in conventional magnetic recording systems, contemporary technology is now moving into optical or optomagnetic systems as candidates for high density E-DRAW (erasable direct read after writing) devices. These devices could be the next generation of information storage devices(1). While these newer systems are superior to conventional magnetic recording systems, the operational principle is based on a thermally induced phase change such as crystal - amorphous transition. Namely, these processes utilize laser heating so that it is a heat-mode device and therefore the merits of optical processes can not be fully appreciated.

Comparison between heat-mode and photon-mode processes is given in Table I. The main differences are the superior resolution and the possibility of multiplex recording in photon-mode systems. Because of the diffusion of heat, the resolution of heat-mode recording is inferior to that of photon-mode systems. Furthermore, photons are rich in information such as energy, polarization and coherency, which can not be rivalled by heat-mode recording.

0097–6156/89/0381–0209$06.00/0
© 1989 American Chemical Society

Table I. Comparison between heat- and photon-mode image recording

item	heat-mode	photon-mode
speed	0	00
sensitivity	0	0
resolution	0	00
area density of image	0	00
storage stability	0	?
read-out stability	0	?
erasability	0	?
rewritability	0	?

0: performance of present heat-mode system, 00: better than heat
mode, ?: questionable, For heat-mode and photon-mode systems,
thermal phase change and conventional photochromic systems are
assumed, respectively.

On the other hand, the heat-mode recording is advantageous in
view of having an energy threshold of recording. Because of this,
the recorded information is not lost after repeated read-out by
monitoring with light of reduced intensity. In contrast, direct
read-out of photochemical(i.e. photochromic) recording at the
wavelength of the photochromic absorption band causes fading of the
recorded image. Photochromic reactions have no energy threshold.
For heat-mode recording, the energy threshold and subsequent phase
change will provide high resolution and high contrast in imaging. In
a light spot from a semiconductor laser, the intensity is the highest
at the center and gradually decreases towards the periphery. If the
threshold energy can be adjusted to the light intensity at the
center, a sharp spot with a diminished size could be recorded even if
the irradiated spot is broad and diffused.
    We are now presenting a new approach to combine the merits of
both heat-mode and photon-mode recordings. Photochemically triggered
phase transition is the basic concept(2, 3). Any molecular aggregate
system can reveal phase transition phenomenon by external or internal
stimuli such as temperature, pressure and chemical composition. When
a molecular aggregate system close to its phase transition condition
is perturbed by a small photochemical change, a phase change can be
triggered and the physical properties will be suddenly altered.
    Expected merits of the photochemically triggered phase transi-
tion system are as follows: Firstly, since the overall changes are
spontaneous once it is triggered, the resultant changes in physical
properties are greatly enhanced. Secondly, the photochemical information is transferred to different physical properties and consequently, the read-out of information can be conducted by some other method
than measuring the photochromic change directly. Thirdly, the phase
transition is a reversible process and the erase-and-rewrite cycle is
ensured. Lastly, the fraction of photochromic change necessary to
induce a phase transition is very small and therefore the fatigue
phenomena common to photochromic compounds can be greatly reduced.
    A shortcoming is the instability against external conditions,
in particular temperature. To induce a phase transition with a minimum amount of photochemical change, the molecular aggregate system

should be placed close to the phase transition condition.   Consequently, the operative temperature range is rather limited.   In other words, the requirements to satisfy a high sensitivity and a wide latitude of temperature are conflicting.

In the following, a number of examples are presented with emphasis on how such amplified imaging device can be built in self-supporting polymer materials.   The present photochromic compounds (azobenzene and spiropyran derivatives) are rather common and have been used many times in molecular assemblies without touching upon the concept of image amplification.   The most recent references are given(4 - 6).

Examples of Photochemically Triggered Physical Amplification

The examples we have demonstrated so far are schematically shown in Figure 1.   Although the photochemical events and the subsequent physical changes are different from system to system, a general trend is the there is a non-linear response to the degree of photochemical reaction, that is, the image amplification has been demonstrated.

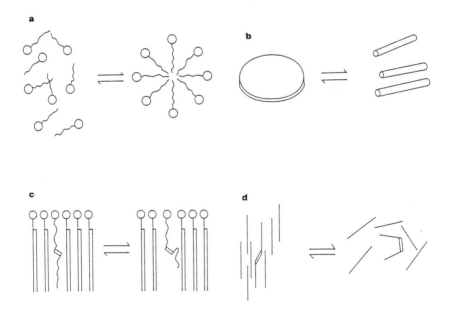

Figure 1. Examples of photochemically triggered physical amplification.
a) Spherical micelle, read-out by change in surface tension. b) Plate-like micelle, read-out by light scattering. c) Vesicle, read-out by circular dichroism. d) Liquid crystals, read-out by polarized light.

Various methods are used for read-out.   Micelle formation and dissociation may be detected by means of a fluorescence probe detect-

ing hydrophobicity. Phase transition in liquid crystals can be
electrically detected. Many other phase change phenomena such as
solid-liquid transitions(crystal melting, solubility change and so
forth), liquid-liquid(homogeneous-phase separation, various phase
transitions in liquid crystals) appear unsuitable for practical imaging materials.

Photochemically Triggered Micelle Formation

A spiropyran compound bearing a pyridinium group and a long alkyl
chain behaves as a surfactant. The components shown in Scheme 1 exhibit reverse photochromism in polar solvents. The colored
merocyanine form is more stable than the spiropyran form in the dark.
Upon photoirradiation at $\lambda > 510$ nm, the polar merocyanine form is con-
verted to the hydrophobic spiropyran form so that the CMC (critical
micelle concentration) of the surfactant decreases. Consequently,
when the initial concentration is set between the CMC of the two
forms, photoirradiation induces a sudden formation of micelles at a
certain conversion to the spiropyran form corresponding to the CMC of
the mixed micelle of the two forms.
     Although CMC is not a sharp phase transition and the degree of
amplification is not phenomenal, a clear non-linear change in surface
tension against the amount of photochemical change is observed as
shown in Figure 2.
     Changes in the shape of the absorption spectrum correspond very
well with micelle formation. The ratio of absorbance at 550 nm to
that at 500 nm(both are absorptions of merocyanine) is constant below
the CMC whereas the value increases continuously with concentration
above CMC. This indicates that the merocyanine is a sensitive probe
to detect micelle formation. During the photoirradiation experiment
shown in Figure 2, the ratio of absorbance started to increase at the
A_t/A_0 value where the surface tension showed a sudden drop.
     When the initial concentration of the merocyanine form is lower
than the CMC of the spiropyran form, the change in surface tension is
gradual all through the progression of photoreaction. The value of
A_{550}/A_{500} remains constant during photoirradiation. Unfortunately,
reversibility of this photochromism is poor and the micelle
formation/dissociation cycle deteriorates rapidly.

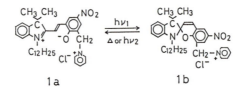

Scheme 1

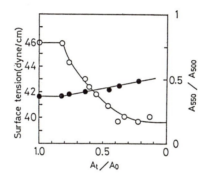

Figure 2. Change in surface tension(○) and absorbance ratio(●) as a function of the degree of photoisomerization(A_t/A_0) of 1a in water. [1] = 5.2×10^{-5} M; A_0 and A_t are absorbance of 1a at 502 nm at time 0 and t, respectively.

Photochemical Control of Aggregation Number

The amphipathic compounds shown in Scheme 2 can form a disc-like
micelle($\underline{7}$). The shape of a molecular aggregate depends on the shape
of the constituent molecules($\underline{8}$). For instance, conical molecules
with large polar head groups prefer to form spherical micelles while
cylindrical molecules tend to give flat aggregates. Trans-
azobenzene is a rod-like molecule whereas the cis-form is bent.

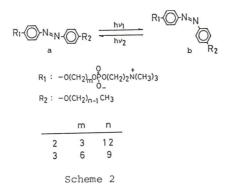

Scheme 2

Consequently, a photochemical transformation from the trans to the
cis isomer changes the state of molecular aggregation. The bent
structure of the cis form cannot be accommodated in the large disc-
like aggregate and thus the aggregation number decreases non-linearly
with the progression of photoisomerization. This change is clearly
shown by the change in light scattering intensity.

The results of differential scanning calorimetry(DSC) indicate
the change in aggregation state. The trans micelle showed a main endothermic peak at $14.2\,^{\circ}C(\Delta H = 1.0\ \text{kcal/mol})$, corresponding to a gel-
liquid crystal phase transition, whereas the transition temperature
for the cis micelle appeared at $11.9\,^{\circ}C(\Delta H = 0.8\ \text{kcal/mol})$. This is
unequivocal evidence that the trans-cis photoisomerization is a sufficient perturbation to alter the state of molecular aggregation.

As shown in Figure 3, the change in R_θ is non-linearly related
to the cis content expressed by the increase in absorbance at 450 nm,
a characteristic absorption band of cis-azobenzene. This nonlinearity is attributed to the intrinsic nature of phase transformation occurring at a critical condition. A sudden change in R_θ com-
mences when the cis content reaches the critical value which can
trigger the transformation.

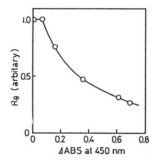

Figure 3. Change in light scattering intensity(R_θ) upon photo-irradiation of the micelle of 2a. [2] = 5.1×10^{-4} M. $\triangle$ABS at 450 nm corresponds to the formation of 2b.

Photochemically Triggered Induced Circular Dichroism in Liposomes
When an optically inactive chromophore is subject to the effect of
optically active environment, optical activity may be induced at the
absorption wavelength of the optically inactive chromophore. This
phenomenon of induced circular dichroism(ICD) is often observed in
polypeptides bearing various achiral chromophores on the side
chain(9). The strong chiral environment caused by the peptide helix
structure is responsible for this. Distance from, and orientation
to, the chiral field decide the degree of ICD appearing on the
achiral chromophore.

Provided that an optically active molecular aggregate is photochemically perturbed to change the state of molecular alignment, the
effect of a chiral environment on an achiral chromophore incorporated
in the molecular aggregate will be also altered. It has been known
that polypeptides bearing photochromic side groups change their optically active properties as a result of photochromic reaction(10-12).
This phenomenon is likely to be related to non-linear
photoresponsiveness.

We have demonstrated that a chiral vesicle composed of di-
palmitoyl-L- α-phosphatidylcholine(l-DPPC) doped with the azobenzene
containing amphiphiles shown in Scheme 2 is a subject to photochemically triggered phase transition and exhibits a non-linear photoresponse in terms of ICD appearing at the absorption band of
azobenzene.

The mixed liposomal solutions were prepared by the ethanol-
injection method(13) in order to obtain completely transparent solutions. It is interesting to note that miscibility of the
photochromic amphiphiles with DPPC depend on the position of bulky
azobenzene. If azobenzene is incorporated close to the end of long
alkyl chain, a stable mixed bilayer state cannot be formed. On the
other hand, when the azobenzene moiety is located near the head group
or at the center of the hydrocarbon tail, the azobenzene amphiphiles
are successfully incorporated into the bilayer membrane. No individual micelle formation nor phase separation in the bilayer was
observed at 25 °C by absorption spectroscopy. However, the
microstructure of the mixed liposomes depends on the type of azobenzene amphiphiles.

A study by DSC provides clear evidence that a homogeneous mixture of DPPC with the photochromic amphiphiles is formed when azobenzene is located at the center of the alkyl chain. The endothermic
peak at 40.9 °C due to gel-liquid crystalline phase transition(T_m) of
DPPC shifts to lower temperature on mixing with 3, the position being
dependent on the amount of the doped azobenzene amphiphiles. When
azobenzene is close to the polar head group(2), the mixed liposomes
is molecularly inhomogeneous. Two peaks in the DSC corresponding to
each component remain unchanged, indicating phase separation.

Photochemical response of these liposomes is different from
each other. With progression of trans - cis photoisomerization of
azobenzene, ICD at the absorption band of the trans isomer decreases.
As shown in Figure 4, depression in ICD is almost proportional to the
amount of photoisomerization for the phase separated system.
Photoisomerization in the domain of azobenzene aggregate proceeds independently from the rest of DPPC aggregate so that the depression
in ICD corresponds to the concentration of remaining trans-
azobenzene. When the two components are molecularly mixed, change of

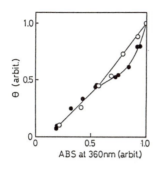

Figure 4. Change in ICD(Θ) upon photoirradiation of 1-DPPC bilayer containing 2(O) and 3($\bullet$). [2 or 3] = 5×10^{-5} M, [1-DPPC] = 5×10^{-4} M.

molecular shape from the rod-like trans form to the bent cis form seems to influence the molecular arrangement in the DPPC liposomes and therefore the chiral environment as well. The amount of ICD depression is larger than that expected from the decrease in the trans azobenzene concentration.

Incidentally, ICD in DPPC liposomes is observed in the temperature range below T_m but not above. Consequently, the non-linear depression of ICD will be relevant to disordering of DPPC molecular arrangement. The change in ICD is a reversible process. Reverse photoisomerization to the trans isomer restores the initial ICD.

Image Amplification by Means of Photochemically Triggered Phase Transition in Liquid Crystal

Although the examples described so far involve image amplification mechanisms, the non-linear response brought about by photochemically induced phase transition is not sharp because of the gradual nature of the phase transitions employed. For example, while CMC is a transition, micelle formation occurs via preliminary molecular association(premicelle formation) and thus the transition is not sharp. Certainly, better defined transitions such as crystal-liquid and crystal-amorphous are thermodynamically more unequivocal in comparison with micelle formation. Although micelles and liquid crystals are phenomenologically similar and indeed, plate-like micelles may be considered to be a special case of lyotropic smectic liquid crystals, phase transitions in thermotropic liquid crystals are better defined and much sharper than the changes occurring at CMC of a micelle.

The use of photoreactive liquid crystal systems in imaging devices is not unprecedented. As early as 1971, Sackmann showed for the first time the photochemical change of pitch in cholesteric liquid crystals (14). Since then several techniques for using liquid crystals in imaging have been reported, i.e., high contrast photoimage with nematic liquid crystals(15), photoimaging under biased potential(16), cholesteric compound tagged with azobenzene(17), and etc. None of them, however, described the concept of image amplification.

4-Cyano-4'-n-pentylbiphenyl(5CB) which formed nematic liquid crystals was doped with 4-butyl-4'-methoxyazobenzene(BMAB) and placed in a thin layer glass cell after surface alignment by rubbing treatment. The sample was irradiated at 355 nm to conduct trans - cis photoisomerization of BMAB. The phase transition induced by the photoisomerization was followed by monitoring at 633 nm(a He-Ne laser) via two crossed polarizers, the sample being placed between them. The strongest monitor signal was obtained when the angle of the monitor light to the cell was   45°. The results(3) are shown in Figure 5. While photoisomerization proceeds nearly linearly with reaction time, the change in monitor signal intensity is drastic.

Depression of the nematic-isotropic phase transition temperature(T_{NI}) is caused by the addition of cis-BMAB. Sudden phase transition occurs when the content of cis isomer reaches the critical

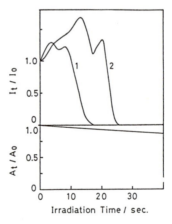

Figure 5. Read-out intensity change(I_t/I_0) and absorbance change (A_t/A_0) owing to trans - cis photoisomerization of BMAB in 5CB liquid crystal at 34°C. [BMAB] = 4.9 mol%(1) and 3.0 mol%(2).

concentration at the particular temperature. This indicates that the sensitivity is strongly dependent on the operating temperature. When 4.9 mol% and 3.0 mol% of trans-BMAB are added, T_{NI} are 36.7°C and

35.6°C, respectively. Irradiation at 34°C brings about a quick response whereas a longer time of irradiation is required at lower temperature.

Evaluation of image amplification may be made by comparing the optical density change(I_t/I_0) with the change in absorbance(A_t/A_0) owing to photochromism of azobenzene. The underlying principle is as follows. When a signal in the form of transmitted light is provided, the sensitivity is decided by the signal-to-noise(S/N) ratio. Since the signal is monitored by a photomultiplier or a pin photodiode, the larger the optical density change per unit input photoenergy the higher the S/N ratio and consequently the sensitivity is higher. The ratio, $\Delta (I_t/I_0)/\Delta (A_t/A_0)$ in a certain period of irradiation represents the degree of amplification with a fixed S/N ratio. The degree of amplification well exceeds 100 under an optimum condition.

The relation between the type of photochromic compound and its effectiveness to induce a phase transition is a point of interest. When unsubstituted azobenzene is added to 5CB, the phase transition is not induced even after prolonged irradiation. BMAB is by itself liquid crystalline whereas azobenzene is not. It seems to be essential for a triggering photochromic compound to have effective interactions with the host liquid crystal.

Erasing of the image can be achieved by switching the photoirradiation to 525 nm to induce cis - trans isomerization of azobenzene. Since the absorbance of the cis isomer at 525 nm is weak, it takes a longer period than the image recording process. Also there seems to be a certain time delay between photoreaction and complete recovery of the nematic phase. This problem is relevant to molecular mobility in liquid crystals as a function of temperature, rubbing condition, external electric field and most importantly, the type of liquid crystal. Research is now being undertaken on direct determination of molecular mobility by fluorescence technique.

Electrical read-out of a photoimage is also possible. A nematic - isotropic phase change disorganizes the arrangement of dipoles and hence the dielectric constant changes. Viscosity is also affected so that the frequency dispersion of dielectric constant is different between nematic and isotropic phases. A condenser was constructed by introducing the photosensitive liquid crystal mixture between two transparent conductive electrodes(ITO glass) separated by 7 μm. Variation of capacitance due to nematic - isotropic phase transition was followed by a capacitance bridge as shown in Figure 6. At 0.1 KHz, the capacitance difference between two phases is the largest. It is rather disappointing that the optimum frequency is so low. A quick response of electric signal is not possible in this system. This situation may be improved by the use of ferroelectric liquid crystals.

Until now, liquid crystals have been used to display an electric signal as a visual pattern. The present demonstration may open a new possibility of the reverse use of liquid crystals, that is, conversion of a photosignal to an electric signal.

In view of sensitivity, the liquid crystal system is much improved than the previously mentioned systems. However, these liquid

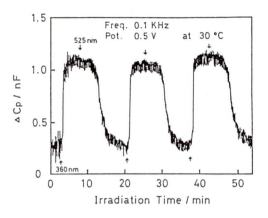

Figure 6. Photochemically induced capacitance change of 5CB at frequency of 0.1 KHz and bias potential 0.5 V at 30 °C. [BMAB] = 5 mol%.

crystalline materials are viscous fluids and thus the long term image stability is not expected. To overcome this shortcoming, image amplification in a solid system has to be designed.

Photochemically Triggered Phase Transition in Liquid Crystalline Polymer Films

Research on liquid crystalline polymers(LCP) is a fashionable subject with the goal of developing speciality polymers of superior mechanical and thermal properties. Besides these properties, other interesting properties of LCP have not been fully utilized. We are trying to use thermotropic LCP for photon-mode image recording material. From the previous demonstration of various phase transitions in small molecular systems, photochemical image recording on polymer films with amplification seems to be a promising approach to a new information storage material. While use of a polymer film will improve image stability when the polymer is kept below T_g, the restricted molecular motion in the solid polymer may reduce the response time.

As a first attempt, we chose a polyacrylate with liquid crystalline side chains as shown in Figure 7. The family of this polymer with different length alkyl spacers has been prepared by Ringsdorf and coworkers(18). While the monomer model compound(i.e. the acrylate before polymerization) does not provide a liquid crystalline phase, and only the crystal - isotropic transition is observed, the polymer shows a clear transition nematic - isotropic transition at ca. 61 °C and the glass transition temperature at 24 °C as shown in Figure 7. T_{NI} depends very much on the length of alkyl spacer. In comparison with the results of Ringsdorf, there seems to be an odd-even effect, which is now under investigation.

The polymer was dissolved in chloroform and doped with 5 mol% of BMAB. The solution was cast on a glass plate and dried to give a film. The sample was subject to monochromatic irradiation at 366 nm at a temperature between T_g and T_{NI} to induce trans-cis photoisomerization of BMAB. The read-out intensity via crossed polarizers is plotted against irradiation time in Figure 8.

A photochemically triggered phase transition is again clearly demonstrated. The apparent increase of the transmittance before its sharp decline is seemingly due to a subtle change in interference of

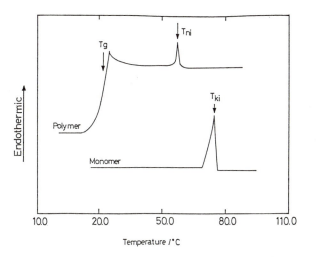

Figure 7. DSC Thermograms of liquid crystalline polymer and its monomer.

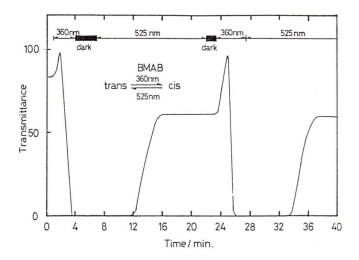

Figure 8. Photochemically triggered phase transition in solid polymer film.

monitoring light. By switching the irradiation wavelength to 525 nm
which is exclusively absorbed by the cis isomer, recovery of trans
BMAB accompanies the restoration of the nematic phase. To induce the
phase transition, the required amount of photoisomerization of BMAB
is extremely small if the operating temperature is close to T_{NI} so
that deterioration of the chromophore during erase-rewrite cycles is
considerably suppressed. When BMAB is replaced by unsubstituted
azobenzene, photoresponse is poor.

Long term storage of an image will be possible for this polymer
system. At room temperature, below the T_g, the recorded information
remains unchanged for many days. It is a dilemma that a phase transition is possible only when molecules can move, while molecular motion blurs the recorded image. Many years ago, one of the authors
presented a concept of image fixation by cooling(19). The concept
was demonstrated by photodimerization of anthracene derivatives
bonded to a polymer. Photodimerization can proceed only above the
temperature somewhat higher than T_g. Depending upon the type of
reaction, the required free volume is different. Consequently, cooling of the system below the critical temperature below which the
available free volume is not sufficient for the photoreaction to
proceed is a handy way of image fixation. In other words, this is a
combination use of heat- and photon-mode recording.

LCP is a suitable candidate for this purpose. Since heat-mode
image recording on LCP has been known(20), the merits of this system
may be enhanced by the aid of a photochemical trigger. As shown in
Figure 9, the photoresponse is strongly temperature dependent.
Preliminary heating close to the phase transition temperature
facilitates the subsequent photochemical imaging with possible high
resolution in comparison with overall heat-mode recording. When the
system is cooled down below the threshold temperature, the image is
stabilized regardless of the state of photochromic molecule. Thermal

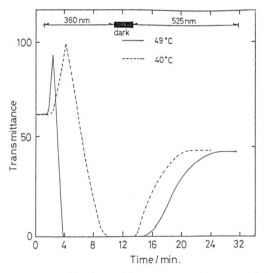

Figure 9. Temperature effect on photoresponsiveness of polymer film.

back reaction of photochromic compound does not affect the frozen-in image in the immobile hard matrix.

Conclusion

Photochemically induced phase transition is a wide-spread phenomenon in molecular aggregate systems in general. An imminent problem to be solved will be how to compromise image recording speed with image stability. Since molecular mobility has an opposite effect on each requirement, some trick to promote or retard molecular motion such as subsidiary heating/cooling cycle will be necessary for the photon-mode phase change materials to be of practical use.

Acknowledgments

This study was supported by Special Coordination Funds of the Science and Technology Agency of the Japanese Government as well as by Grant-in-Aid for Special Study #61123002.

Literature Cited

1   Technical Digest of papers presented at International Symposium on Optical Memory 1987, Tokyo, September 1987.
2   Tazuke, S.; Kurihara, S.; Yamaguchi, H.; Ikeda,T. J. Phys. Chem. , 1987, 91, 249.
3   Tazuke, S.; Kurihara, S.; Ikeda, T. Chem. Lett. 1987, 911.
4   Suzuki, Y.: Ozawa, K.; Hosoki, A.; Ichimura, K. Polym. Bull. 1987, 17, 285.
5   Seki, T.; Ichimura, K. Chem. Comm. 1987, 1189.
6   Ramesh, V.; Labes, M. M. J. Am. Chem. Soc. 1987, 109, 3228.
7   Okahata, Y.; Ihara, H.; Shimomura, M.; Tamaki, S.; Kunitake, T. Chem. Lett. 1980, 1169.
8   Israelachivili, J. N. Intermolecular and Surface Forces; Academic: New York, 1985; Chapter 15.
9   Hatano, M, Adv. Polym. Sci. 986, 77, 66.
10   Pieroni, O.; Houben, J. L.; Fissi, A.; Costantino, P.; Ciardelli, F. J. Am. Chem. Soc. 1980, 102, 5913.
11   Ueno, A.; Takahashi, K.; Anzai, J.; Osa, T. J. Am. Chem. Soc. 1981, 103, 6410.
12   Yamamoto, H.; Macromolecules, 1986, 19, 2472.
13   Batzri, S.; Korn, E. D. Biochim. Biophys. Acta. 1973, 298, 1015.
14   Sackmann, E.; J. Am. Chem.Soc. 1971, 93, 7088.
15   Haas, W. E.; Nelson, K. F.; Adams, J. E.; Dir, G. A. J. Electrochem. Soc. 1974, 121, 1667.
16   Ogura, K.; Hirabayashi, H.; Uejimas, A.; Nakamura, K. Jpn. J. Appl. Phys. 1982, 21, 969.
17   Irie, M.; Shiode, Y.; Hayashi, K. Polym. Preprints Jpn. 1986, 35, 487.
18   Portugall, M.; Ringsdorf, H.; Zentel, R. Makromol. Chem. 1982, 183, 2311.
19   Tazuke, S.; Hayashi, N. J. Polym. Sci., Polym. Chem. Ed. 1978, 16, 2729.
20   Eich, M.; Wendorff, J. H.; Ringsdorf, H. Makromol. Chem., Rapid Commun. 1987, 8, 59.

RECEIVED September 6, 1988

Chapter 14

Radiation Durability of Polymeric Matrix Composites

Darrel R. Tenney and Wayne S. Slemp

Materials Division, Langley Research Center, National Air and Space Administration, Hampton, VA 23665–5225

Experimental results are presented that show that high doses of electron radiation combined with thermal cycling can significantly change the mechanical and physical properties of graphite fiber-reinforced polymer-matrix composites. Polymeric materials examined have included 121°C and 177°C cure epoxies, polyimide, amorphous thermoplastic, and semicrystalline thermoplastics. Composite panels fabricated and tested included: four-ply unidirectional, four-ply [0, 90, 90, 0] and eight-ply quasi-isotropic [0/±45/90]s. Test specimens with fiber orientations of [10] and [45] were cut from the unidirectional panels to determine shear properties. Mechanical and physical property tests were conducted at cold (-157°C), room (24°C) and elevated (121°C) temperatures. Tests results show that electron radiation significantly lowers the glass transition temperature of graphite/epoxy composites. Radiation degradation products plasticized the matrix at elevated temperatures and embrittled the matrix at low temperatures. Low-temperature matrix embrittlement resulted in a significant increase in the number of microcracks formed in the composites during thermal cycling. Matrix microcracking changed the coefficient of thermal expansion of the composites. In general, electron radiation degraded the tensile and shear properties of the composite at both cold and elevated temperatures. Compression properties were improved at low temperatures but degraded at elevated temperatures. Results from tests on neat resins showed a good correlation between neat resins and composite response.

The use of graphite fiber reinforced polymer-matrix composites for structural applications on weight critical space structures has greatly increased in recent years. The main advantage of composites over conventional aerospace metallic materials is their high specific strength and stiffness and low thermal expansion. High-stiffness, low-expansion graphite/epoxy composites have been used extensively on communication satellites (1) (over 40,000 parts in orbit on INTEL SAT V communication satellites) and composites are considered the leading candidate materials for nearly every space

This chapter not subject to U.S. copyright
Published 1989 American Chemical Society

structure under development or being considered. This includes applications as diverse as long slender composite tubes (15-m long x 5-cm diameter) for large structures such as the truss structure for Space Station (2) and lightweight high-precision composite panels (hexagonal honeycomb core/composite face sheets panels 2m on a side which weigh 6-10 Kg/m^2 and have a surface roughness < 2 μm rms) for precision segmented reflectors (3) currently being developed as part of NASA's Astronomical Observatory Program.

The effect of the space environment on composites (4) is an important consideration in determining the design lifetime of a space structure. Spacecraft placed in low earth orbit (i.e., Space Station, 500 km) will be subjected to high vacuum, thermal cycling (every 90 min, -157 to 121°C, worst case), UV radiation, atomic oxygen, and possible micrometeoroid and space debris impact. The composite tubes being developed for the Space Station truss structure are covered with aluminum foil (.05 mm) to protect the composite from UV radiation and atomic oxygen erosion. Also, the exterior aluminum surface will be anodized to give a solar absorptance to emittance ratio of 0.30/0.65 to minimize the thermal cycle. Because these tubes will be subjected to approximately 175,000 thermal cycles during the design lifetime (30 years) of Space Station, an extensive testing program is currently underway to evaluate the effect of repeated thermal cycles on composite properties (5). Tests are also being conducted to characterize hypervelocity impact damage to composite tubes, panels, and filament wound pressure vessels being considered for Space Station.

Spacecraft placed in geosynchronous orbit (GEO) will be subjected to high vacuum, UV radiation, proton and electron radiation (4). Ultraviolet radiation and low-energy electron and proton radiation will generally be stopped in coatings placed on composite structures for thermal control and environmental protection. However, high energy (1 - 10 MeV) electron radiation can easily penetrate the entire thickness of a typical 4- to 8-ply composite laminate resulting in nearly uniform energy deposition through the thickness of the composite. A total cumulative dose reaching approximately 1×10^{10} rads is expected for a structure at GEO for 25-30 years. A dose of 10^9 rads is generally considered to be the threshold for significant property changes in polymeric materials (6). Therefore doses in excess of 10^9 rads would be expected to affect the mechanical and physical properties of polymer matrix composites.

During the past five years research has been conducted at NASA Langley Research Center to determine the effect of high-energy electron radiation on the mechanical and physical properties of polymer-matrix composites. Materials examined have included graphite-reinforced epoxies, polyimides, and thermoplastics. Extensive testing was performed over a broad temperature range (-157 to 121°C) to determine the magnitude of property changes due to radiation exposure and irradiation followed by thermal cycling. Analytical characterization was also performed to determine damage mechanisms to guide new materials development. In-depth reporting of the results of these studies can be found in references (7-20). The results presented in this paper represent an assimilation of the key findings of this work.

EXPERIMENTAL

Materials. A summary of the composite materials included in the radiation effects studies conducted at NASA Langley Research Center during the past 6 years is given in Table I. All of the graphite/epoxy composites were fabricated using T300 fiber. The PAN-based fibers C6000 and AS4 used for the polyimide and thermoplastic composites have similar properties to the T300 fibers used for the epoxy composites. All materials except the two semicrystalline thermoplastics were produced from

Table I. Composite Materials Studied in Radiation Exposures

Composite Type	Composite Fiber[1]/Matrix	Prepreg Source	Processing	Matrix Description
Graphite/ Epoxy	T300/CE339	Ferro Corp	LaRC[2]	121°C cure 2-phase elastomer toughened epoxy
	T300/CE339 (Mod.)	Ferro Corp	LaRC	121°C cure epoxy; no elastomer addit.
	T300/5208	NARMCO	LaRC	177°C cure, MY720 based epoxy
	T300/934	Fiberite Corp.	LaRC	177°C cure, MY720 based epoxy
	T300/934 (Mod.)	Fiberite Corp.	LaRC	177°C cure, MY720 based epoxy; process. additive removed
	T300/BP907	American Cyanamid	LaRC	177°C cure, single phase toughened epoxy
Graphite/ Polyimide	C6000/PMR15	U.S. Polymer	LaRC	315°C cure
Graphite/ Thermo- plastic	C6000/P1700	Union Carbide Corp.	LaRC	Polysulfone, amorphous thermoplastic
	AS4/PPS	Phillips Petroleum	Phillips Petroleum	Polyphenylene sulfide, semicrystalline thermoplastic
	AS4/PEEK	ICI Americas Inc.	ICI	Polyetherether ketone, semicrystalline thermoplastic

[1] T300, C6000, and AS4 are intermediate modules (30-40 msi) PAN-based graphite fibers
[2] Processed at the Langley Research Center using manufacturers' recommended cure cycle

unidirectional prepreg tape purchased from the manufacturers indicated in Table I. Composite panel layup and processing were conducted at NASA LaRC using manufacturers' recommended cure procedures. Composite panels fabricated included: 4-ply unidirectional, 4-ply [0,90,90,0] and 8-ply quasi-isotropic [0/±45/90]$_s$. Individual specimens were machined from these panels for exposure to radiation, and/or thermal cycling, and subsequent characterization. Specimens for tensile testing were 1.27 cm wide and 15.25 cm long. Specimens for dynamic mechanical analyses (DMA) and thermomechanical analysis (TMA) were 3.8 cm by 1.2 cm and 0.62 cm by 0.62 cm respectively. To avoid effects due to moisture all specimens were dried for 30 days at room temperature in vacuum as a preconditioning step in the experimental program.

Radiation Exposure. A Radiation Dynamics Corporation Dynamatron accelerator was used to expose the specimens to 1-MeV electrons in a turbopumped vacuum chamber at a pressure of 2×10^{-7} torr. Up to 19 specimens were mounted side by side on a temperature-controlled aluminum mounting plate positioned in the uniform area of the electron beam (approx. 25-cm in diameter) and perpendicular to the incoming electrons. Faraday cups mounted in the exposure area of the base plate were used to monitor the electron flux. A maximum dose rate of 5×10^7 rad/h was used to keep the specimen temperature below 38°C during the exposures. The maximum total dose given any of the specimens was 1×10^{10} rads.

Mechanical Property Testing. Mechanical tests were performed on both unirradiated and irradiated materials at -157°C, 24°C, and 121°C. Specimens were kept dry prior to testing in an environmental chamber mounted in a tensile testing machine. Tensile test specimens of [0]$_4$, [10]$_4$, [45]$_4$, and [90]$_4$ laminates were cut from 4-ply composite panels. All specimens were straight-sided coupons. For tension and shear tests the length/width aspect ratio was 8. For the compression tests the aspect ratio was 0.25 and the unsupported length was 0.64 cm. The [0]$_4$ laminates were used to measure the ultimate tension and compression strength, X_1, the axial modulus, E_1, and Poisson's ratio, ν_{12}, (fiber direction). Transverse properties Y_T and E_2 were determined from [90]$_4$ laminates. Shear strength was calculated from [10]$_4$ laminates and the shear modulus, G_{12}, from the [45]$_4$ laminates. Strain was measured using strain gages and fiberglass tabs were used for load introduction.

Dynamic Mechanical and Thermomechanical Analysis. A DuPont Model 981 DMA was used to determine the dynamic modulus and damping characteristics of baseline and irradiated specimens. Transverse composite samples 1.27 cm x 2.5 cm were used so that the modulus and damping data were primarily sensitive to matrix effects. Data were generally determined from -120°C through the glass transition temperature (T_g) of each material using a heating rate of 5°C/min.

Laminate expansion/contraction characteristics were determined with a DuPont Model 942 TMA. In this test the movement of an 0.38 cm diameter, hemispherical-tipped quartz probe resting on the specimen was monitored as the specimen was heated from room temperature through its softening range.

Thermal Expansion Measurement. Thermal expansion measurements were made with a laser interferometer dilatometer (20) with a strain resolution of approximately 2×10^{-6}. The temperature cycle for all tests went from room temperature to a maximum of 121°C (except where noted), down to -157°C and back to room temperature. Thermal strain data were taken at approximately 20°C increments with a 30-minute hold at each temperature to allow the specimen and interferometer to reach thermal equilibrium. Specimens were dried prior to testing and a dry N_2 purge was used to keep the specimen dry while in the interferometer.

RADIATION EFFECTS ON THERMAL PROPERTIES

Electron irradiation causes chain scission and crosslinking in polymers. Both of these phenomena directly affect the glass transition temperature (T_g) of the materials. Thermomechanical (TMA) and dynamic-mechanical analysis (DMA) provide information about the T_g region and its changes due to radiation damage. Therefore, DMA and TMA were performed on all irradiated materials.

Typical DMA results ([7]) for a 177°C cure Gr/Epoxy system (T300/934) are shown in Figure 1. The damping data show that irradiation results in: (1) lowering T_g about 50°C, (2) increasing the height of the T_g peak, and (3) significantly broadening the T_g peak of the [90°] specimen. The lowering of T_g and the higher damping peaks indicate that irradiation has lowered the molecular weight between crosslinks and lowered crosslink density of the epoxy matrix. The broad damping peak for the 90° specimen indicates that irradiation also widens the distribution of molecular weights. Another indication of changing the molecular weight between crosslinks is the very broad "rubbery region" (i.e., region under the damping peak as defined in Figure 1) of the irradiated sample compared to that of the unirradiated composite sample. The broad "rubbery region" reflects the presence of irradiation-induced degradation products that plasticize the matrix and thereby enhance the damping compared to that of the unirradiated material. These changes would be expected to affect the matrix-dominated mechanical properties and thermal expansion characteristics of the composites.

The formation of low molecular weight products in irradiated 934 epoxy composites was confirmed by performing TMA and vacuum weight loss studies ([7]). Typical results are shown in Figures 2 and 3. The TMA curves in Figure 2 show that for the irradiated sample, softening begins near 100°C and significant expansion/swelling occurs near 150°C. The observed expansion is believed due to the volatilization of trapped radiation-induced degradation products. Blistering and delamination were noted on some samples heated to 170°C in separate tests.

The vacuum-thermal weight loss results presented in Figure 3 show a two-step weight loss pattern for the irradiated sample. Weight loss associated with volatilization of the first product begins before 100°C and the second near 150°C. No weight loss was observed below 225°C for the nonirradiated composite sample.

Typical DMA and TMA results ([8]) for a 121°C cure rubber toughened composite system (T300/CE339) after different levels of radiation exposure are shown in Figures 4-7. The glass transition (115°C) and secondary relaxation temperatures (-50°C) of the baseline epoxy matrix and the apparent T_g of the elastomer (-30°C) were altered by the irradiation exposures. The "rubbery range" was expanded nearly 50°C in the low temperature direction. The expansion of the "rubbery range" as measured by half-width of the T_g peak at half peak height is shown in Figure 5. The threshold for significant change in the T_g region is 10^8 rads. Broadening of the T_g region indicates that the epoxy network structure is degraded by irradiation. However, radiation-induced crosslinking must also be part of the interaction mechanism because the location of T_g did not shift to lower temperatures as observed for the 177°C cure epoxy systems.

The damping characteristics of the composite in the low temperature region are shown in Figure 6. The secondary relaxation peak of the epoxy is shifted to lower temperatures with higher levels of irradiation. This shift indicates matrix degradation because less energy is required for the short-range molecular motion in the irradiated material. The electron exposures also changed the structure of the

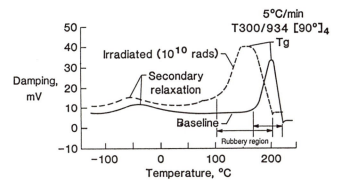

Figure 1. Dynamic mechanical analysis of T300/934 composite. (Reproduced from reference 7.)

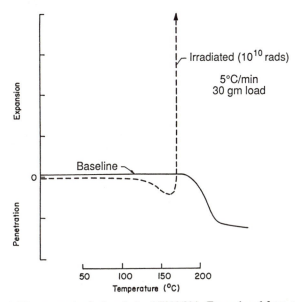

Figure 2. Thermomechanical analysis of T300/934. (Reproduced from reference 7.)

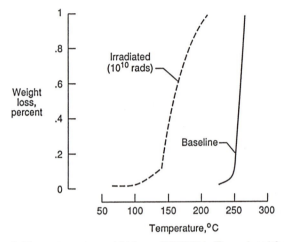

Figure 3. Thermal vacuum weight loss of T300/934. (Reproduced from reference 7.)

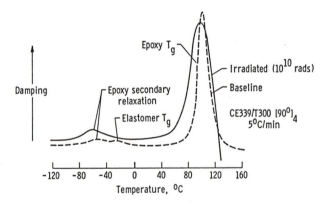

Figure 4. Dynamic mechanical analysis of T300/CE 339. (Reproduced from reference 8.)

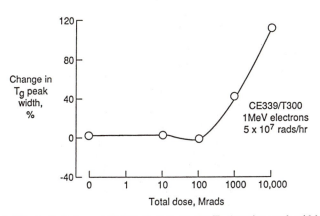

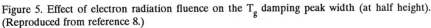

Figure 5. Effect of electron radiation fluence on the T_g damping peak width (at half height). (Reproduced from reference 8.)

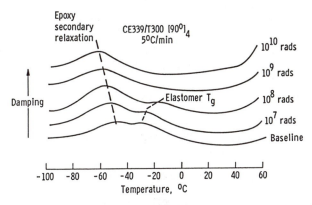

Figure 6. Dynamic mechanical damping of CE 339/T300 at low temperatures. (Reproduced from reference 8.)

CTBN type elastomers (carboxyl-terminated butadiene acrylonitrile) used to toughen this epoxy system. The T_g peak of the elastomer (-30°C) was shifted to higher temperatures suggesting that electron interaction resulted in crosslinking. The fact that the T_g peak disappeared at 10^9 rads suggests that crosslinking in the elastomer was extensive at high dose levels.

TMA data for T300/CE339 are shown in Figure 7. The relatively low temperature softening of the irradiated laminates compared to the baseline laminate suggest matrix network degradation. The lower rate of softening (slope of the penetration curve) of the irradiated laminates at higher temperatures suggests the presence of a crosslinked network. For the highest level of exposure examined (10^{10} rads) either expansion or penetration was observed as the sample was heated near 100°C. Delamination and blistering believed to have been caused by volatilized degradation products which could not escape resulted in expansion for some samples. For other samples where volatilized products were able to escape, possibly through microcracks in the matrix, a penetration curve was observed similar to that found for specimens receiving lower total doses.

Thermal analyses were also performed on thermoplastic composites studied at NASA (9). DMA and TMA results for three Gr/epoxy and three Gr/thermoplastic composites are compared in Figures 8 and 9. The secondary relaxation peaks of the unirradiated composites in the -150°C to 100°C temperature range are shown in Figure 8. Figure 9 shows the effect of thermal cycling only, and irradiation followed by thermal cycling on the matrix glass transition temperature and softening associated with this transition for each of the composite materials. All specimens received 500 thermal cycles between -150°C and an upper temperature of 65°C (T300/CE339 and AS4/PPS) or 93°C (T300/934, C6000/P1700, AS4/PEEK, T300/BP907). In all cases the upper temperature was at least 25°C below the T_g of the composite.

The results in Figure 9 show that thermal cycling and radiation change the materials. For the two semicrystalline thermoplastic composites (AS4/PPS and AS4/PEEK) the degree of softening as measured by TMA penetration was significantly reduced by thermal cycling indicating that these materials were annealed during the thermal cycling and may have increased in crystallinity. Thermal cycling of irradiated composites produced changes in the T_g and related softening range of all of the composites. The T_g of the two 177°C cure epoxies (934 and BP907) was significantly reduced and the softening range was extended over a broader temperature range. Also, the T_g of the amorphous thermoplastic (C6000/P1700) was similarly reduced indicating that electron radiation degrades the matrix. The effects of these changes on mechanical and thermal expansion properties will be discussed in subsequent sections of this paper.

RADIATION EFFECTS ON MECHANICAL PROPERTIES

During the past 6 years research has been conducted at NASA Langley Research Center to determine the effect of high energy electron radiation exposure on the mechanical properties of the composites listed in Table I. The bulk of this work has been on the T300/934 composite system. Tests on this system have included: static tension, compression, and shear at -157°C, RT, and 121°C (10-13); cyclic tension and shears tests at -157°C, RT, and 121°C (14-15); interlaminar fracture toughness tests at -157°C, RT, and 121°C (16). Modulus and strength properties of the composites are summarized in Table II and resin tensile properties are summarized in Table III. Tensile tests were also conducted for the T300/CE339 system at -115°C, RT, and 80°C (8). Tensile and interlaminar fracture toughness testing was performed for the T300/5208 system at RT (17). Tensile tests were also conducted on T300/934, T300/BP907, T300/CE339, C6000/P1700, AS4/PPS, and AS4/PEEK

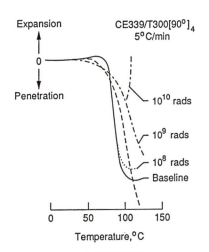

Figure 7. Thermomechanical analysis of T300/CE 339. (Reproduced from reference 8.)

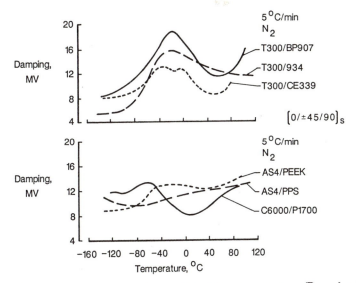

Figure 8. Dynamic mechanical damping of composites at low temperatures. (Reproduced from reference 9.)

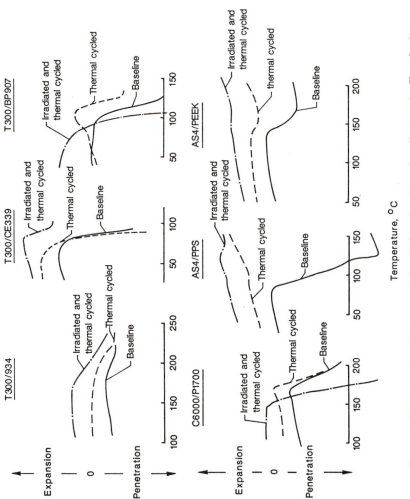

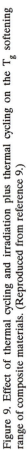

Figure 9. Effect of thermal cycling and irradiation plus thermal cycling on the T_g softening range of composite materials. (Reproduced from reference 9.)

Table II.   Tensile and Compressive Properties of T300/934

Temp	Prop	Baseline Tension*	Δ_RT%	Compression	Δ_RT%	Irradiated Tension*	Δ_RT%	Compression	Δ_RT%	Δ_IR% Tens.*	Comp.
-157°C	X	141	-36	241	25	127	-43	228	21	10	-5
	Y	4.56	-51	56.5	86	2.81	-60	51.8	97	-38	-8
	S	7.34	-26	---	---	7.25	-22	---	---	-1.2	---
	E_1	18.6	-1.6	17.2	11	19.2	n.c.	16.0	10	3.2	-7
	E_2	1.83	33	2.10	29	2.12	39	2.14	23	16	2
	v_{12}	0.313	n.c.	0.136	-25	0.368	30	0.104	-32	18	-24
	v_{21}	---	---	0.0320	-13	---	---	0.0324	-28	---	1
	G_{12}	1.17	70	---	---	1.12	44	---	---	-4.3	---
Room	X	222	---	193	---	223	---	188	---	n.c.	-3
	Y	9.37	---	30.4	---	6.98	---	26.3	---	-26	-13
	S	9.92	---	---	---	9.25	---	---	---	-6.8	---
	E_1	18.9	---	15.5	---	19.3	---	14.6	---	2.1	-6
	E_2	1.38	---	1.63	---	1.52	---	1.74	---	10	7
	v_{12}	0.314	---	0.182	---	0.283	---	0.153	---	-10	-16
	v_{21}	---	---	0.0366	---	---	---	0.0447	---	---	22
	G_{12}	0.688	---	---	---	0.777	---	---	---	13	---
121°C	X	194	-13	126	-35	162	-27	48.3	-74	-16	-62
	Y	6.76	-28	18.9	-37	5.88	-16	8.71	-67	-13	-54
	S	5.97	-40	---	---	4.06	-56	---	---	-32	---
	E_1	19.0	n.c.	15.0	-3	19.8	2.6	14.4	-1	4.2	-4
	E_2	1.24	-10	1.80	10	1.06	-30	0.666	-62	-15	-63
	v_{12}	0.345	9.9	0.309	70	0.397	40	0.384	151	15	24
	v_{21}	---	---	0.0791	116	---	---	0.239	435	---	237
	G_{12}	0.563	-18	---	---	0.397	-49	---	---	-29	---

All properties in U.S. Customary units (e.g. strengths in ksi, moduli in msi, and strains in percent)

$\Delta_{RT}\%$ percent change from room temperature value
$\Delta_{IR}\%$ percent change of irradiated w.r.t baseline value
* Tensile properties the results of Milkovich, et al. (1984)

Table III. Fiberite 934 Tensile Properties

Temp	Prop		Baseline Value	%Δ_{RT}	Irradiated Value	%Δ_{RT}	%Δ_{IR}
Room	E	(msi)	0.674	---	0.799	---	+19
	v		0.363	---	0.373	---	+3
	G	(msi)	0.247	---	0.291	---	+18
	σ_{PL}	(ksi)	3.42	---	5.15	---	+51
	σ_{ult}	(ksi)	8.53	---	10.9	---	+28
	ε_{ult}	(%)	1.48	---	1.55	---	+5
121°C	E	(msi)	0.489	-27	0.501	-37	+2
	v		0.341	-6	0.368	-1	+7
	G	(msi)	0.182	-26	0.183	-37	n.c.
	σ_{PL}	(ksi)	2.44	-29	1.34	-74	-45
	σ_{max}	(ksi)	10.3	+21	6.61	-39	-36
	ε_{max}	(%)	3.05	+106	2.54	+64	-17

Δ_{RT}% percent change from room temperature value
Δ_{IR}% percent change of irradiated w.r.t baseline value

after irradiation and irradiation followed by 500 thermal cycles (9). An overview of selected parts of this work is presented in the following sections.

Tension. Tensile stress strain curves for baseline and irradiated 934 neat resin (13) are shown in Figure 10. Irradiation resulted in an increase in the strength and stiffness at room temperature. However, at elevated test temperature the strength and stiffness were both reduced by irradiation. The strength of the baseline resin was 21% higher at elevated temperatures than the room temperature value but the irradiated sample had an elevated temperature strength 39% below the room temperature value. At high temperatures low molecular by products of irradiation plasticize the matrix, resulting in the nonlinear stress-strain behavior shown in Figure 10.

Typical tension stress-strain curves of baseline and irradiated unidirectional T300/934 composites tested in [0] and [90] orientations at three different temperatures (12) are shown in Figures 11 and 12. Irradiation had essentially no effect on the fiber-dominated tensile modulus of the [0] specimen and caused only a small (10-15%) reduction in strength at the low and elevated temperatures. For the matrix-dominated [90] laminates, irradiation caused a very substantial decrease in strength at three test temperatures (-38% at -157°C, -26% R.T., -13% 121°C). Irradiation increased the modulus at -157°C and R.T. (10 - 15%), but lowered it at 121°C (-15%). These results are consistent with results obtained on the neat resin specimens discussed above.

Shear. The axial tensile responses of [10] and [45]of axis specimens (12) are shown in Figures 13 and 14. For the [10] specimen irradiation caused a large decrease in the shear modulus (-29%) and a large decrease in shear strength (-32%) at the elevated temperature. Irradiation had only a small effect on the properties at -157°C and R.T. For the [45] specimen radiation similarly reduced the elevated temperature modulus and strength but increased the strain to failure. At 121°C the strain to failure of the [45] specimen (Figure 14) was nearly double that of the [90] specimen (Figure 12). Radiation-induced damage appears to plasticize the matrix and increase the strain to failure of the matrix in shear but reduce the transverse tensile strain to failure of the [90] specimen possibly due to a reduction in the fiber/matrix bond strength which controls failure.

Compression. The effect of irradiation on the axial and transverse compressive response of T300/934 unidirectional composite (13) is shown in Figures 15 and 16 and in Table II. Irradiation had very little effect on the compressive properties at -157°C and caused only a small reduction in the strength properties at room temperature, -3% axial and -13% transverse. However, irradiation caused a severe reduction in the strength of both the [0] and [90] laminates (-74% and -67%, respectively) at elevated temperatures. The elevated temperature stress-strain curve for the [90] laminate in Figure 16 clearly shows that the matrix has been degraded by irradiation which explains the reduced strength of the [0] specimen at elevated temperatures shown in Figure 15. At room and cold temperatures the matrix stiffness is sufficient to prevent microbuckling of the fibers such that the strength of the composite reflects the strength of the fibers. However, at elevated temperatures the matrix stiffness is reduced to the point where lateral support for the fibers is not sufficient to achieve full fiber properties. These results are consistent with results presented in Table III for neat resin specimens tested at elevated temperatures with and without irradiation exposure. The DMA results for baseline and irradiated T300/934 presented in Figure 1 showed that the average molecular weight and crosslink density of this material was reduced by irradiation. Both of these effects would be expected to reduce the elevated temperature stiffness of the resin and thus degrade the compressive properties of the composite.

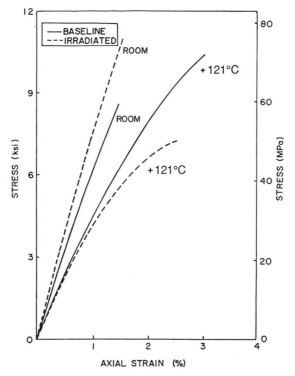

Figure 10. Neat resin tensile response of 934 resin specimens. (Reproduced from reference 13.)

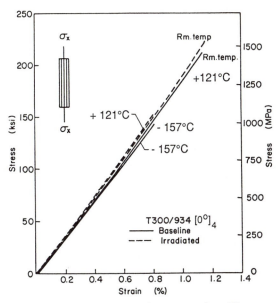

Figure 11. Temperature-dependent stress–strain curves for [0] composite laminates. (Reproduced from reference 12.)

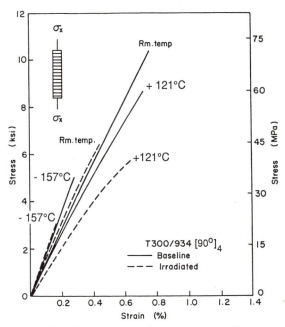

Figure 12. Temperature-dependent stress–strain curves for [90] composite laminates. (Reproduced from reference 12.)

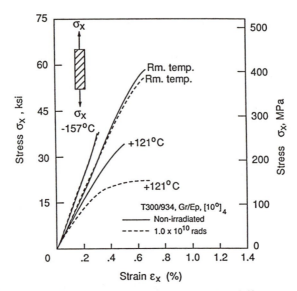

Figure 13. Temperature-dependent stress–strain curves for [10] composite laminates. (Reproduced from reference 12.)

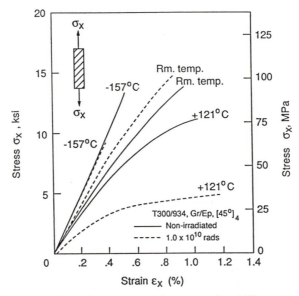

Figure 14. Temperature-dependent stress–strain curves for [45] composite laminates. (Reproduced from reference 12.)

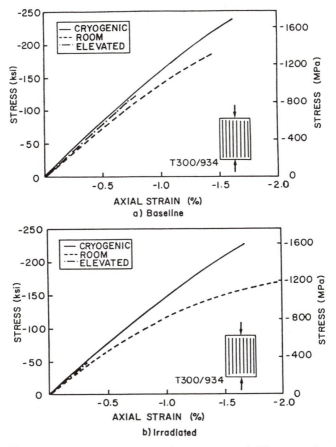

Figure 15. Temperature-dependent compressive response of [0] composite laminates. (Reproduced from reference 13.)

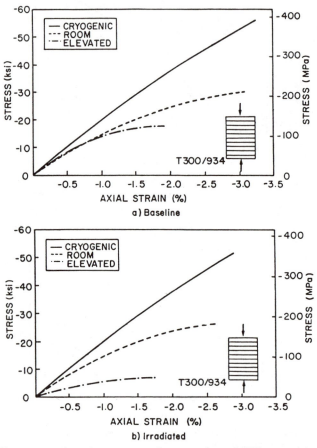

Figure 16. Temperature-dependent compressive response of [90] composite laminates. (Reproduced from reference 13.)

Radiation and Thermal Cycling. An experimental program was conducted (9) to determine the combined effects of radiation and thermal cycling on the mechanical properties of three Gr/epoxy and three Gr/thermoplastic composites. Tension tests were performed on 8-ply quasi-isotropic $[0/\pm 45/90]_s$ laminates in the baseline condition, after radiation exposure to 10^{10} rads (1 MeV electrons), and after radiation exposure followed by 500 thermal cycles between -150°C and an upper temperature of either 65°C (T300/CE339 and AS4/PPS) or 93°C (T300/934, C6000/P1700, AS4/PEEK, T300/BP907). These temperatures were at least 25°C below the Tg of the composites. All cycling was done in dry nitrogen and tensile tests were conducted at room temperature.

The effects of thermal cycling and radiation plus thermal cycling on the modulus and strength of these laminates are shown in Figures 17 and 18.

The data show the tensile modulus (0° direction) and the tensile stress at delamination onset or ultimate stress if no delamination occurred. In general, 500 thermal cycles caused significant matrix microcracking in four of the six materials tested, Figure 19, but did not significantly change either the tensile modulus or strength properties. However, the combination of radiation and thermal cycling significantly degraded the strength properties in all of the laminates. This strength reduction is believed to be due to degradation of the matrix resins and the development of matrix microcracks, Figure 19. BP907 and CE339 were severely degraded as evidenced by extensive matrix microcracking and delamination after irradiation and thermal cycling.

The elastic properties of the epoxies were also reduced by the combined exposure conditions. The large reductions measured for BP907 and CE339 composites are believed due to the presence of delaminations in the materials prior to mechanical testing. The elastic modulus of C6000/P1700 and AS4/PEEK increased after the combined exposure to radiation and thermal cycling. The magnitude of the increase was surprising in view of the changes which occurred in the other composites.

RADIATION EFFECTS ON DIMENSIONAL STABILITY

Radiation degradation of matrix resins can affect the dimensional stability of polymer matrix composites in two ways. Radiation-induced chain scission can produce degradation products that plasticize the matrix at elevated temperatures which can change the way in which residual curing stresses are relieved in the composite, and degradation products can embrittle the matrix at low temperatures resulting in matrix microcracking. Irradiation can also result in additional cross-linking which can embrittle the matrix resin. Both of these changes will affect the thermal expansion behavior of composites.

Figure 20 shows the effects of radiation degradation products on the thermal expansion behavior (18) of a typical 177°C cure Gr/Ep composite ($[0_2/90_2]_s$ T300/5208). The irradiated specimen shows a pronounced nonlinearity at elevated temperature and a permanent negative residual strain of approximately -67×10^{-6} at room temperature after one thermal cycle to -150°C. Strains of this magnitude would significantly affect the performance of dimensionally stable space structures. Repeated cycles over the same temperature range give a strain response parallel to the unirradiated curve but displaced by the permanent residual strain present after the first cycle. However, if the specimen was cycled to a higher maximum temperature an additional change in slope of the thermal strain curve occurs which results in an additional permanent residual strain. However, the coefficient of thermal

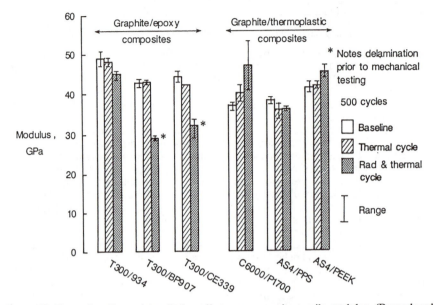

Figure 17. Thermal cycling and irradiation effects on composite tensile modulus. (Reproduced from reference 9.)

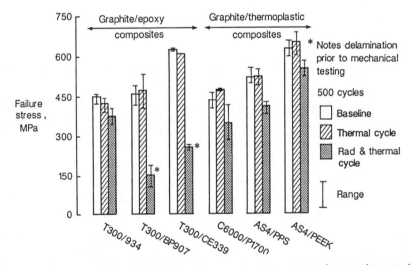

Figure 18. Thermal cycling and irradiation effects on the failure stress of composite materials. (Reproduced from reference 9.)

Microcrack density, cracks/cm

	Baseline	Thermal cycled	Irradiated and thermal cycled [1]
T300/934	0	7	17
T300/BP907	0	0	> 50 [2]
T300/CE339	0	8	25
C6000/P1700	5	21	24
AS4/PPS	0	19	22
AS4/PEEK	0	1	5

1 Irradiated to 10^{10} rads at 5×10^7 rads/hr followed by 500 thermal cycles

2 Cracking and delamination extensive

Figure 19. Effect of thermal cycling and irradiation plus thermal cycling on microdamage development in composite materials. (Reproduced from reference 9.)

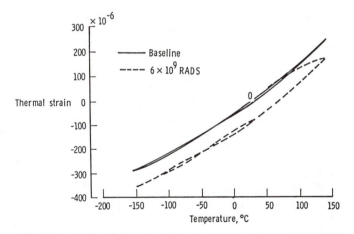

Figure 20. Effects of radiation on the thermal expansion of T300/5208 graphite/epoxy composite. (Reproduced from reference 18.)

expansion (CTE) of radiated and unirradiated specimens was the same at all temperatures except when a nonlinear behavior was observed at the highest temperature investigated.

The nonlinear strain response and subsequent permanent residual strain at room temperature is related to radiation degradation products plasticizing the matrix. This can best be explained by the DMA results (18) presented in Figure 21. The damping data for the irradiated composite show that the T_g is lowered by approximately 22°C and a broad "rubbery region" is produced compared to the nonirradiated composite sample. During the thermal cycling tests the specimen was heated into the region where the matrix could flow thus relieving residual tensile curing stresses resulting in a more fiber dominated response at high temperature (nonlinear region) and permanent negative residual strains at room temperature. The reason for no additional changes on subsequent heating cycles may be related to the procedure used to run the thermal expansion tests. The heating process in these tests occurred slowly in 22°C increments, with 30 minute holds at each temperature. In the 107°-138°C temperature range, chemical changes apparently took place resulting in a movement of the "rubbery region" back to higher temperatures out of the thermal expansion test range.

Tests of a graphite-reinforced polyimide composite (C6000/PMR15) did not show any effect of radiation exposure (1 MEV electrons 6×10^9 rad total dose) on the thermal expansion behavior (14). DMA curves for unirradiated and irradiated composites were essentially identical over the temperature range of the thermal expansion measurements.

Matrix microcracking can significantly change the CTE of the composite laminate (4) as illustrated in Figure 22. Matrix microcracking can cause as much as a 70-80 percent change in CTE of a $[0_2/90_2]_s$ laminate. Radiation exposure to 10^{10} rads followed by 500 thermal cycles between -156°C and 121°C has been shown to result in a crack density of approximately 30 cracks/cm compared to only ~7-8 cracks/cm for thermal cycling only. Thus radiation products which embrittle the resin at low temperatures can result in higher levels of matrix microcracking and significantly change the CTE response of composite laminates.

The crack densities developed in a rubber-toughened 121°C cure Gr/Ep composite (T300/CE339) are shown (8) in Figure 23. Samples were exposed to 1 MEV electrons for total doses ranging from 10 to 10,000 Mrads and then cycled 1, 10, and 100 times between -115°C and 80°C. No microcracks were found in the unexposed composite even after 100 thermal cycles, showing that this epoxy system is more thermal fatigue resistant than 177°C cure epoxy systems. However, after irradiation all samples microcracked with a rapid increase in damage above approximately 10^8 rads of exposure. These results are consistent with the changes noted in the DMA and TMA curves for this system noted in Figures 4-7. The threshold for significant changes in the T_g region was 10^8 rads (Figure 5) which is also the threshold for a rapid increase in microcracking density due to thermal cycling, Figure 23. Matrix embrittlement at low temperatures was due to formation of low molecular weight products due to radiation degradation of the epoxy network and embrittlement of the elastomer (CTBN) through a crosslinking mechanism, Figure 6. The embrittlement associated with both of these changes will be expected to degrade the thermal fatigue properties of the composites.

The effect of radiation on the thermal expansion of this toughened composite (T300/CE 339) is shown (19) in Figure 24. The thermal strains measured during the cool-down portion of the first thermal cycle (cooling from RT to -150°C) are shown for the baseline composite (no radiation exposure) and for samples exposed to total doses as high as 10^{10} rads. Radiation levels, as low as 10^8 rads

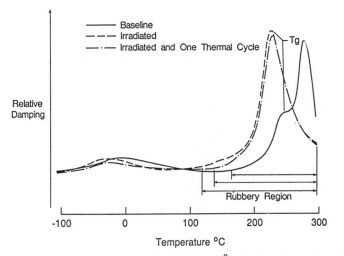

Figure 21. Effects of 1 MeV electron radiation (6×10^9 rads fluence) on the relative damping of $[0/\pm45/90]_s$ T300/5208 graphite/epoxy composite. (Reproduced from reference 18.)

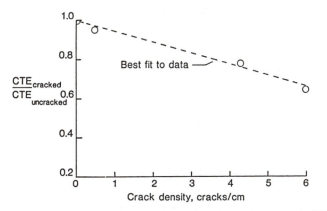

Figure 22. Effect of crack density on coefficient of thermal expansion of a T300/5208 composite laminate $[0_2/90_2]_s$. (Reproduced from reference 4.)

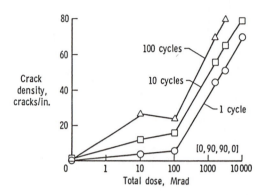

Figure 23. Effect of radiation and thermal cycling between −115 °C and 80 °C on microcrack density. (Reproduced from reference 8.)

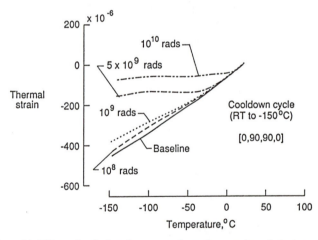

Figure 24. Effect of radiation fluence on thermal expansion of elastomer-toughened specimens during cool down from room temperature. (Reproduced from reference 19.)

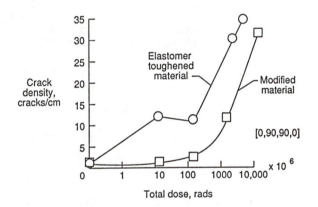

Figure 25. Effect of radiation fluence on microcrack formation in composite specimens subjected to 100 thermal cycles between 80 °C and −150 °C. (Reproduced from reference 19.)

(equivalent to about 1 year in GEO) changed the expansion characteristics of the resins. Major reductions in CTE are observed for doses above 10^9 rads due to extensive microcracking of the matrix.

To assess the effect of elastomer degradation on composite performance, additional composites were fabricated with the same 121°C cure epoxy without any addition of the elastomer (21). The expansion behavior of the modified epoxy composite was similar to the toughened material. For electron doses less than 10^9 rads the CTE of the toughened and untoughened composites were essentially the same which suggests that the epoxy matrix and not the elastomeric component controls the thermal expansion behavior.

A comparison of the effect of radiation on thermal cycling damage development in the toughened and untoughened composites is shown (19) in Figure 25. Removing the elastomer shifted the microcracking curve to higher total doses indicating that the (CTBN) elastomer is radiation sensitive and contributes to composite embrittlement at low radiation dose levels. However, for high radiation doses, above 10^8 rads, extensive microcracking of the base epoxy occurred. Because of the severe matrix embrittlement which occurs at doses above 10^8 rads, neither T300/CE339 or the modified T300/CE339 would be acceptable for long-term (20–25 years) service in geosynchronous earth orbit where the accumulated total dose could be well in excess of 10^9 rads.

CONCLUDING REMARKS

High doses (>10^9 rads) of electron irradiation, typical of doses expected in composite structural members of spacecraft located at GEO for 20-30 years, can significantly alter the chemical, physical, and mechanical properties of polymeric matrix composites. Radiation was observed to alter the chemical structure of the polymers by both chain scission and crosslinking. The formation of low molecular weight products in the polymer plasticized the matrix at elevated temperatures and embrittled the matrix at low temperatures. Plasticization of the matrix at elevated temperatures significantly reduced the shear and compression properties of the composites and resulted in a permanent residual strain development in cross-ply laminates upon exposure to elevated temperature (into the rubbery region). Embrittlement of the matrix at low temperatures resulted in enhanced matrix microcracking during thermal cycling after radiation exposure. Matrix microcracks lower the coefficient of thermal expansion of the composite and at high crack densities (>20 cracks/cm) can degrade the strengths and stiffness properties of the laminates. Low-temperature curing epoxies (i.e., 121°C) were found to be particularly susceptible to radiation damage and are judged to lack the durability required for long-life space missions in the Earth's trapped radiation belts. High temperature polyimides and thermoplastics, such as PEEK, appear to have very good radiation resistance, but may not be acceptable for high precision space structures because of high residual stresses in the composites due to their high processing temperatures. Overall, the 171°C cure epoxies appear to offer good potential for long-life space structures.

LITERATURE CITED

1.   Hillesland, H.L. SAMPE Symposium/Exhibition, 1980, vol. 24, pp. 202-211
2.   Mikulas, M.M., et al. NASA TM-86338, 1985.
3.   Swanson, P. N. Jet Propulsion Laboratory JPL D-2283, June 1985.
4.   Tenney, D. R.; Sykes, G. F.; and Bowles, D. E. Proc. Third European Symp. Spacecraft Matls in Space Environment, Oct 1985, ESA SP-232, pp. 9-21.
5.   Tompkins, S. S.; Bowles, D. E.; Slemp, W. S.; and Teichman, L. A. Space Station Symposium, AIAA Paper No. 88-2476, April 1988.

6. Kircher, J. F.; and Bowman, R. E. Effects of Radiation on Materials and Components. Reinhold Publ. Co., New York, 1964.
7. Sykes, G. F.; Milkovich, S. M.; and Herakovich, C. T. Polymeric Materials Science and Engineering, Proc. of the ACS Div. Polymeric Matls and Engrg., vol. 52, ACS, 1985, pp. 598-603.
8. Sykes, G. F.; and Slemp, W. S. Proceedings of the 30th SAMPE National Symposium-Exhibition, Anaheim, Ca, March 19-21, 1985; pp. 1356-1368.
9. Sykes, G. F.; Funk, J. G.; and Slemp, W. S. Proceedings of the 18th International SAMPE Technical Conference, Seattle, WA, October 7-9, 1986; pp. 520-534.
10. Milkovich, S. M.; Herakovich, C. T.; and Sykes, G. F. The NASA-Virginia Tech Composites Program Interim Report 45, VPI-E-84-20, June 1984.
11. Milkovich, S. M.; Sykes, G. F.: and Herakovich, C. T. Fractography of Modern Engineering Materials--Composites and Metals, ASTM STP 948, Nov. 1985, pp. 217-237.
12. Milkovich, S. M.; Herakovich, C. T. and Sykes, G. F. J. Comp. Matls, 20, 6, 1986, pp. 579-593.
13. Fox, D. J.; Herakovich, C. T.; and Sykes, G. F. The NASA-Virginia Tech Composites Program Interim Report 63, also CCMS 87-11 and VPI-E-87, July 1987.
14. Reed, S. M.; Herakovich, C. T.; and Sykes, G. F. The NASA-Virginia Tech Composites Program Interim Report 60, VPI-E-86-19, October 1986.p
15. Reed, S. M.; Sykes, G. F.; and Herakovich, C. T.   J. Reinforced Plastics and Composites, July 1987, 6, 3, pp. 234-252.
16. Funk, J. G.; and Sykes, G. F. J. Comp. Tech. & Research, Fall 1986, 8, 3, pp. 92-97.
17. Wolf, K. W.; Fornes, R. E.; Gilbert, R. D. and Memory, J. D. J. Appl. Phys. Sci., Oct. 1983, 54, 10, pp. 5558-5561.
18. Bowles, D. E.; Tompkins, S. S.; and Sykes, G. F. AIAA J. Spacecraft and Rockets, Nov-Dec 1986, 23, 6, pp. 625-629.
19. Sykes, G. F.; and Bowles, D. E. SAMPE Qrtly, July 1986, 17, 4, pp. 39-45.
20. Tompkins, S. S.; Bowles, D. E.; and Kennedy, W. R.: SEM Experimental Mechanics, Mar. 1986, 26, 1, pp. 1-6.

RECEIVED September 12, 1988

Chapter 15

Radiation-Resistant, Amorphous, All-Aromatic Poly(arylene ether sulfones)

Synthesis, Physical Behavior, and Degradation Characteristics

D. A. Lewis[1,3], James H. O'Donnell[1], J. L. Hedrick[2,4], T. C. Ward[2], and J. E. McGrath[2,5]

[1]Polymer and Radiation Group, Department of Chemistry, University of Queensland, St. Lucia, Brisbane 4067, Australia
[2]Polymer Materials and Interfaces Laboratory, Department of Chemistry, Virginia Polytechnic Institute and State University, Blacksburg, VA 24061

An aromatic polysulfone based on 4,4'-biphenol and 4,4'-dichlorodiphenyl sulfone (Bp PSF) was shown to be the most resistant of a systematic series of poly (arylene ether sulfones) to ^{60}Co gamma irradiation. Sulfur dioxide was the major volatile product and was used as a probe to correlate the radiation resistance with polymer structure. The use of biphenol in the polymer reduced G(SO_2) by 60% compared with bisphenol A based systems (Bis-A PSF). Surprisingly, the isopropylidene group was shown to be remarkably radiation resistant. The ultimate tensile strain decreased with dose for all polysulfones investigated and the rate of decrease correlated well with the order of radiation resistance determined from volatile product measurements. The fracture toughness (K_{IC}) of Bis-A PSF also decreased with irradiation dose, but the biphenol based system maintained its original ductility.

There is an increasing need for the production of light, very strong polymeric based matrix resins, adhesives and composite structures which can withstand harsh environments such as UV and ionizing radiation. High performance engineering thermoplastics such as aromatic polysulfones and polyether ketones are strong candidates for this application since they are easy to process, have high maximum service temperature, good modulus and fracture toughness characteristics. An extended service life is required to render such structures economically viable, thus it is necessary

[3]Current address: Watson Research Center, IBM Corporation, Yorktown Heights, NY 10598
[4]Current address: Almaden Research Center, IBM Corporation, San Jose, CA 95120–6099
[5]Address correspondence to this author.

0097–6156/89/0381–0252$06.00/0
© 1989 American Chemical Society

to optimize radiation resistance while still maintaining the required mechanical properties. Two aromatic polysulfones which are currently available commercially have been shown (1,2) to be resistant to ionizing radiation, making this class of polymers suitable for engineering applications in these harsh environments.

There have been relatively few systematic studies of the relationship between polymer structure and radiation resistance for polymers with aromatic backbones. Such investigations are necessary to determine which linkages are the most susceptible to radiation degradation. The paucity of data is due, in part, to the difficulties in synthesizing a series of polymers with controlled structure and molecular weight. Thus, the synthesis of the new polymers was an important part of this work. All of the polymers are linear with no long chain branching and are all para substituted. These variables have been shown to have important effects on the radiation resistance. Furthermore, the polymers have similar molecular weights and molecular weight distributions, which allow the comparison of the change in mechanical properties with radiation and polymer structure to be a more defined structure/property investigation.

The production of volatile products upon irradiation can result in the undesirable formation of bubbles. This can lead to the premature failure of stressed components, thus limiting service life. The yield of volatile products after irradiation has been demonstrated to be a sensitive probe of the relative radiation resistance for model compounds such as cyclohexane and benzene, with $G(H_2)$ = 5.6 (3) and 0.038 (4), respectively.

The mechanical properties exhibited by a polymer after irradiation are a complex function of molecular weight and molecular weight distribution and the number and type of new structures formed. Thus, it is difficult to draw structure/radiation resistance conclusions from the change in the mechanical properties alone. However, the changes in mechanical properties are direct indications of the ultimate usefulness of the polymer in a radiation environment.

In this paper, we examine the relationship between radiation resistance and polymer structure using volatile product and mechanical property measurements.

Experimental

The polymers used in this study were prepared by a nucleophilic activated aromatic substitution reaction of a bisphenate and dihalo diphenyl sulfone (5). The reaction was carried out in an aprotic dipolar solvent (NMP) at 170°C in the presence of potassium carbonate (Scheme 1) (5,6). The polymers were purified by repeated precipitation into methanol/water, followed by drying to constant weight. The bisphenols used were bisphenol-A (Bis-A), hydroquinone (Hq) and biphenol (Bp). Thus, the aliphatic character of Bis-A could be removed while retaining a similar aromatic content and structure. The use of biphenol allows an investigation of the possible effect of extended conjugation on the radiation degradation.

Scheme 1

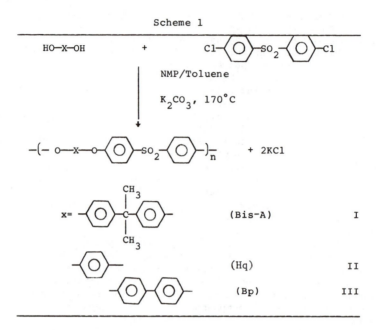

$x=$ (Bis-A)   I

(Hq)   II

(Bp)   III

The poly(arylene ether sulfones) comprise strictly alternating bisphenol and diphenyl sulfone units. Similarly, "copolymers" were prepared using two different bisphenols, which produced a statistical sequence with the alternating diphenyl sulfone units.

The composition and structures of these polymers were characterized by ^{1}H and ^{13}C NMR and infrared spectroscopy to confirm that linear, all <u>para</u> substituted polymers were formed. GPC measurements were used to demonstrate that a monomodal molecular weight distribution with a polydispersity of approximately 2 was obtained. The glass transition temperature was determined on a Perkin-Elmer DSC-2 using a heating rate of 10°C/min and the intrinsic viscosity was determined in NMP at 25°C. These data are summarized in Table I.

The technique used for volatile product analysis was based on the quantitative transfer of volatile small molecules which are produced by radiolysis of the polymer onto a GC column where they are separated, identified and determined quantitatively, as described in detail earlier (<u>7</u>). A weighed sample (ca. 40 mg) of polymer was evacuated and sealed in a thin walled glass ampoule (approx. 20mm x 2mm diam) after slowly heating under vacuum to 200°C over 24 hours to remove adsorbed water. After irradiation, the ampoule was heated in the specially designed injection port (<u>8</u>) of a Hewlett Packard 5730 A gas chromatograph at 150°C for 15 min prior to being crushed by a plunger. The gases were carried from the injection port onto a Chromosorb 102 column (3m x3mm) using helium carrier gas and then to a thermal conductivity detector and

Table I
Polysulfone Characterization

Polymer	$[\eta]_{NMP}^{25^{\circ}C}$	$Tg/^{\circ}C$
Bis-A PSF	0.50	190
Hq PSF	1.40	217
Bp PSF	1.00	232
Hq/Bp(50) PSF	0.71	217

flame ionization detector connected in series. The GC conditions were optimized to give baseline resolution between successive peaks, enabling identification of low molecular weight gases and accurate peak area determination. Calibration plots of peak area versus gas volume were established using pure gases.

Dogbone samples were cut from compression molded thin sheets. They were bundled together in groups of four between glass microscope slides (1 mm thick) to eliminate the effect of non-uniform distribution of secondary electrons (9). Samples for fracture toughness measurements and the dogbone bundles were then sealed in Pyrex tubes after heating under high vacuum to 150°C to remove air and water. ^{60}Co gamma irradiation was conducted with dose rates of approximately 10kGyh^{-1} as determined by extended Fricke Dosimetry. Ampoules for volatile product studies were irradiated at 150±1°C in an aluminum block heater which was controlled by a Eurotherm temperature controller. This temperature was selected since it was near the maximum service temperature for these polymers. In addition, relatively small doses of irradiation (100-600 kGy) were required to obtain easily quantified volumes of gases. Fracture toughness, K_{IC}, measurements were performed according to ASTM standard E399 on specimens 6.3 mm thick at a strain rate of 0.5 mm/min. Stress-strain measurements were made on an Instron tensile tester at a strain rate of 10 mm/min.

Results

The major volatile product from the irradiation of Bis-A PSF at 150°C was sulfur dioxide, which was produced with $G(SO_2) = 0.146$. This is consistent with previous measurements at 30°C, 125°C and 220°C (1,2). Other volatile products observed were hydrogen, methane, and carbon dioxide. The G values for the various gaseous products are compared with literature results for irradiation at 30°C in Table II.

The apparent reduction in $G(H_2)$ at the higher irradiation temperature used in this work is believed to be an artifact due to a reduction in the hyrocarbon impurities in the samples used in this study compared with the work of Brown and O'Donnell (1,2). This is supported by the absence of C_3 hydrocarbons in the volatile products after irradiation at 150°C. Similarly, the water observed in the volatile products after irradiation at ambient temperature

Table II
G(gas) for Bis-A PSF, Irradiated in Vacuum

Product	$30°C^a$	$150°C^b$
SO_2	0.02	0.146 ± 0.014
H_2O	0.009	---
H_2	0.008	0.0033 ± 0.0009
CO	---	0.0050 ± 0.0012
CO_2	0.002	0.0070 ± 0.0019
CH_4	0.002	0.0066 ± 0.0012
C_3 hydrocarbons	0.0002	---
Totals	0.041	0.163

a from Brown and O'Donnell, ref. 1.
b errors are estimated at the 95% confidence level

is believed to be due to a less rigorous drying procedure employed
in that study.

It is unlikely that carbon dioxide would be produced from the
degrading polymer since multiple bond fragmentation and reformation
would be required. After repeated precipitation of the polymer,
$G(CO_2)$ decreased markedly while the yield of other products was not
affected. It is proposed that potassium carbonate, used to
generate reactive phenolate during the step growth polymerization,
was occluded in the precipitated polymer and was the primary source
of the observed carbon dioxide.

Elimination of a methyl radical from the isopropylidene group
of Bis-A PSF, followed by hydrogen abstraction to form methane
might be expected to be an important process, since this is the
only aliphatic part of this polymer. However, $G(CH_4)$ was very
small compared with $G(SO_2)$ and it appears that C-CH$_3$ scission is a
relatively low yield process. This implies that the isopropylidene
group is a relatively radiation resistant group in a structure such
as bisphenol-A. This conclusion is supported by ESR and NMR
studies (10) which demonstrate that main chain scission also does
not occur at this group, at least at neutral pH. The results are
contrary to the conclusions of Sasuga (11) (high dose rate electron
beam irradiation in air) where the radiation resistance of the
isopropylidene group was only slightly greater than the sulfone
linkage. Perhaps the actual irradiation temperature was quite
different in their experiments due to the high dose rates employed.

The dependence of the volatile product yield with structure
can be a very sensitive probe of radiation resistance and the
protective effect of aromatic rings. $G(H_2)$ was observed to
decrease from 5.6 to 0.038 for cyclohexane (3) and benzene (4)
after gamma irradiation at ambient temperature. Since all polymers
under investigation contained the sulfone moiety, $G(SO_2)$ (Table
III) is an ideal probe for radiation resistance for this series.

Table III
G(SO_2) for Several Poly(Arylene Ether
Sulfones) After Vacuum Irradiation at 150°C

Polymer	G(SO_2)[a]
Bis-A PSF	0.146 ± 0.014
Hq PSF	0.136 ± 0.011
Hq/Bp(50) PSF	0.097 ± 0.004
Bp PSF	0.063 ± 0.010

[a] errors correspond to 95% confidence level

The relatively minor role of the isopropylidene group in the radiation degradation of Bis-A PSF was further demonstrated by the small difference in G(SO_2) for Bis-A PSF (0.146) and Hq PSF (0.136), a wholly aromatic polymer. The small reduction in G(SO_2) may be due to a slightly higher aromatic content in Hq PSF compared with Bis-A PSF.

The most radiation resistant polysulfone investigated was Bp PSF, G(SO_2) of 0.063. Although the sulfone group is not directly attached to the biphenyl group, there is evidently a large protective effect either through space or along the chain. This is in accordance with the substantially greater radiation resistance of biphenyl, compared with benzene as shown by G(radical) values of 0.045 and 0.2(12) respectively. The total "aromatic" content of Bp PSF is only slightly higher than for Bis-A PSF and thus does not account for the large increase in radiation resistance observed.

One possibility is that the radical cations and anions formed immediately after irradiation are stabilized to a greater extent in the biphenyl case than for a single phenyl ring. The aromatic rings in the biphenyl group are nearly 90° out of the plane of the molecule, which suggests that there is little interaction between these rings, for the neutral group. After ionization or electron capture (to form the radical cation and radical anion respectively), the orientation of the rings could be altered, which would allow greater interaction and thus charge delocalization. This proposed reorientation clearly is one explanation for the enhanced stability, compared with radical cations or anions derived from aromatic rings such as found in Bis-A PSF or Hq PSF.

G(SO_2) for Hq/Bp (50) PSF is intermediate between G(SO_2) for the homopolymers. Thus, there is no additional protective effect analogous to the non-linear response of G(H_2) as the mole fraction of benzene in cyclohexane is increased (3,4). This indicates that the spatial range of enhanced radiation protection afforded by the biphenyl group is limited.

Ultimately, it is the retention of mechanical properties after irradiation which will determine the suitability of a polymer for use in a radiation environment. Since the potential applications for this class of polymer require high modulus and toughness over

an extended dose range, the properties of most interest are Youngs
modulus, ultimate strain and the fracture toughness.

Youngs modulus increased significantly after a dose of 500
kGy, but increased only slightly more at higher doses, as can be
seen in Table IV. The increase is attributed to the radiation

Table IV
Influence of Irradiation on Mechanical Behavior

	Bis-A PSF		Hq/BP(50) PSF	
Dose/kGy[a]	Modulus/MPa	Ultimate Elongation/%	Modulus/MPa	Ultimate Elongation/%
0	1200	110	1163	74
500	1660	80	1650	100
1000	1750	33	1650	70
2000	1600	20	1750	60
4000	1880	8	1700	51

[a]   Irradiation conducted at 30°C under vacuum

induced network formation and is consistent with observations with
other systems (6,11). The modulus for Hq/Bp (50) PSF increases in a
similar manner, suggesting similarity in the crosslinking reactions
of these two polymers.

The mechanical property which is most sensitive to radiation
degradation is the elongation at failure. This invariably
decreases regardless of whether chain scission or crosslinking is
predominant. For Bis-A PSF, the elongation at failure decreases
rapidly from 110% initially to 8% after a dose of 4000 kGy at
ambient temperature as shown in Table IV. This result is in
remarkable agreement with the electron beam irradiation of Bis-A
PSF in vacuum (6), which showed a reduction in the elongation at
failure from 110% initially to 12% after a radiation dose of 3600
kGy. This is evidence for little dose rate dependency in the
radiation degradation of Bis-A PSF, for dose rates up to 3600
kGyh^{-1} for the film thickness and cooling used in that study.

For Hq/Bp(50) PSF, the decrease in the elongation at failure
with irradiation dose is significantly less than for Bis-A PSF.
This can be attributed to the increased radiation resistance
afforded by the biphenyl moiety, as demonstrated from $G(SO_2)$
measurements.

The decrease in the elongation at failure suggested that the
fracture toughness might also decrease after irradiation. This was
confirmed from the K_{IC} measurements which showed a decrease from an
initial value of 2.0×10^6 Nm$^{-3/2}$ to 1.7×10^6 Nm$^{-3/2}$ after 1000 kGy.
The decrease in K_{IC} correlated with the decrease in tensile
elongation at failure. A final value of 1.4×10^6 Nm$^{-3/2}$ after 4000
kGy was observed.

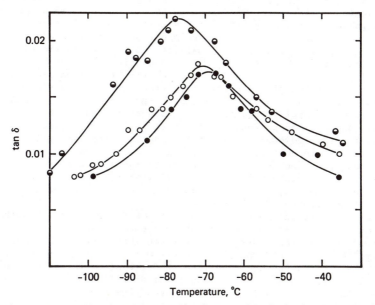

Figure 1. Influence of Irradiation on β-Relaxation of Poly(Arylene Ether Sulfones) Bis-A polysulfone (◐) unirradiated, (O) after 930 kGy and (●) after 4000 kGy.

Fracture toughness may correlate with the β relaxation temperature for the polymer. After irradiation, the β relaxation temperature increases with a corresponding broadening and decrease in intensity which can be seen in Figure 1. This result is consistent with the results of Hinkley et. al. (13) who observed the same phenomenon for polyether sulfone irradiated with electron beam irradiation above Tg.

Conclusions

This research demonstrates the utility of a well-defined set of polymers with carefully controlled structure for relating structure to radiation resistance. The presence of the isopropylidene group in the polymer apparently had little effect on the radiation resistance of the polymer, as determined from volatile product yields, contrary to initial expectations. $G(CH_4)$ was extremely small, indicating that isopropylidene bond scission is of a low probability. This was further confirmed from $G(SO_2)$ measurements.

Bp PSF was the most radiation resistant polysulfone studied. This is attributed to a stabilization of the radical cations and anions formed immediately after absorption of energy by the biphenyl group. The protective effect of this group apparently acts either intramolecularly or through space, since the principal radiation induced volatile product, sulfur dioxide, is derived from the sulfone group which is not adjacent to the biphenyl group. The overall aromatic content of Bp PSF is similar to Bis-A PSF and this difference is not significant enough to explain the enhanced radiation resistance of Bp PSF observed.

The increase in the modulus for Bis A PSF and Hq/Bp PSF with irradiation indicated that crosslinking predominated for both polymers and that the crosslink structures were probably basically similar. Hq/Bp(50) PSF was considerably more radiation resistant than Bis-A PSF, as shown by the rate of decrease in the elongation at failure. For both polymers, there was an initial rapid decrease in the elongation at failure followed by a slower decrease. This effect was also demonstrated by the variation in the fracture toughness (K_{IC}) with irradiation for Bis-A PSF. This work with cobalt-60 gamma radiation complements earlier studies of these materials using high dose rate electron beam irradiation (6).

Acknowledgment

The authors appreciate the support of the NASA Langley Research Center for portions of this research.

Literature Cited

1. Brown, J. R.; O'Donnell, J. H., J. Polym. Sci., Polym Lett., 1970, 8, 121 .
2. Brown, J. R.; O'Donnell, J. H., J. Appl. Poly. Sci., 1975, 19, 405.
3. Ho, S. K.; Freeman, G. R., J. Phys. Chem., 1964, 68, 2189.
4. Gordon, S.; Van Dyken, A. R.; Doumani, T. F., J. Phys. Chem., 1958, 62, 20.

5. Viswanathan, R.; Johnson, B. C.; and McGrath, J. E.; Polymer, 1984, 25, 1827.
6. Hedrick, J. L.; Mohanty, D. K.; Johnson, B. C.; Viswanathan, R.; Hinkley, J. A.; McGrath, J. E., J. Polym. Sci.; Chem. Ed., 1986, 23, 287.
7. Bowmer, T. N.; O'Donnell, J. H., Polymer, 1977, 18, 1032.
8. Bowmer, T. N. Ph.D. Thesis, University of Queensland, Australia, 1979.
9. O'Donnell, J. H.; Sangster, D. F., Principles of Radiation Chemistry, 1970 Edward Arnold, London.
10. Lewis, D. A. Ph.D. Thesis, University of Queensland, Australia, 1988.
11. Sasuga, T.; Hayakawa, N.; Yoshida, K.; Hagiwara, M., Polymer, 1985, 26, 1039.
12. Ayscough, P. B., Electron Spin Resonance in Chemistry, 1967 Butler and Tanner, London, 343.
13. Hinkley, J. A.; J. Polym. Sci.; Polym. Lett. Ed., 1984, 22, 497.

RECEIVED July 29, 1988

Author Index

Affiliation Index

Subject Index

Production and indexing by A. Maureen R. Rouhi
Jacket design by Karen Ruckman

Elements typeset by Hot Type Ltd., Washington, DC
Printed and bound by Maple Press, York, PA

DATE DUE

AUG 1 1 '89 EAS	AUG 2 2 2005
DEC 2 0 '89 EAS	JUL 1 2 2005
ILL sold to JL 12-X90	
MAY 0 7 1993	
MAY 0 7 REC'D	
AUG 0 6 1993	
DEC 1 7 1993	
DEC 2 0 REC'D	
MAY 0 6 1994	
MAY 0 8 REC'D	
MAY 3 0 1995	
MAY 0 9 REC'D	
JUN 1 4 REC'D	
JUN 19 REC'D	
OhioLINK	
DEC 0 9 REC'D	

RET'D

WITHDRAWN

DEMCO 38-297

A113   0937093 7

SCI QD 381.9 .R3 E42 1989

The Effects of radiation on
high-technology polymers

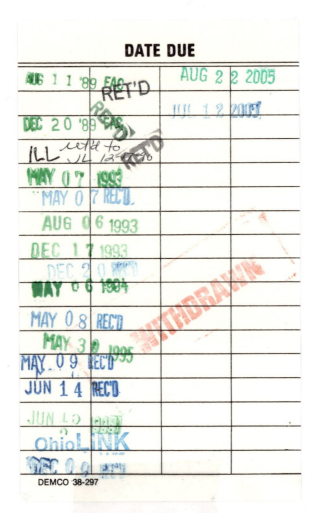